Civil Engineering Drawing

D. V. Jude   Lecturer, Department of Civil Engineering, Loughborough University of Technology

McGRAW-HILL Book Company (UK) Limited

London · New York · St Louis · San Francisco · Auckland · Beirut
Bogotá · Düsseldorf · Johannesburg · Lisbon · Lucerne · Madrid
Mexico · Montreal · New Delhi · Panama · Paris · San Juan · São Paulo
Singapore · Sydney · Tokyo · Toronto

iii

To Karen, Lisa, and Penny
without whose help
this book would have been written
in half the time

Published by
McGraw-Hill Publishing Company Limited
Maidenhead Berkshire England

07 094132 7

Printed and bound in Singapore by FEP International Pte Ltd.

Civil Engineering Drawing

Consulting Editor
Professor P. B. Morice, University of Southampton

Road and rail transport, the Tamar bridges. The new suspension road bridge, built about a century later, beside Brunel's railway bridge. (Photograph supplied by Mott, Hay & Anderson)

Contents

PREFACE

The last quarter of the twentieth century will see more construction than ever before in the history of mankind. More people will require better accommodation, transport systems, and recreational facilities. As civil engineers, we are responsible for the basis of the environment that man has come to accept; we design and build his roads, docks, and airports; we provide him with water and remove his sewage; we also play a large part in providing him with gas and electricity. For this to be done more quickly, more efficiently, and more cheaply than in the past, there must be a constant interchange of ideas and precise information between all engineering disciplines and all nations.

It is here that engineering drawing comes into its own as the undisputed international language of the engineer. A language that must be acquired by the student of civil engineering as soon as possible.

The aim of this book is to teach some of the fundamentals of this language in a straightforward way, for the civil engineering student has, in the past, lacked books dealing specifically with engineering drawing. There are many good books concerned with projection and geometry, as well as others incorporating 'engineering drawing' in the title, but no book known to the author has attempted to provide primarily for the needs of the civil engineer.

The content of all technological courses increases year by year and time must be saved somewhere. This can be done, to some extent, if all the student's exercises, including drawing, are concerned with civil engineering, so that he begins to see and think professionally from the start

Although the later sections of this book, which discuss drawing in practice, may not affect the student immediately, they are intended to give him a background to the role of drawing in civil engineering and should help to explain why things are done in the way they are. It is hoped that the photographs of schemes and structures will play a part in conveying how exciting and challenging the field can be.

SI units *(Système International d'Unités)* have been adopted principally throughout the text, although many of the illustrative examples, taken from British professional practice, were prepared in Imperial units before the changeover to SI.

Several references are made in the text to the necessity of saving time, since skilled men's time is so expensive, and to the need to reduce errors and chances of ambiguity. To this end, the student should familiarize himself at an early stage with the appropriate British Standards, references to which are given in the text.

Acknowledgements

I have received the greatest possible help in the collection of material and facts for this book, from both present and former colleagues, and also from new acquaintances.

The sources of the photographs and the working drawings reproduced facsimile are included in their captions. While gratefully acknowledging permission to use this material, I should also like to thank many other individuals for so willingly giving their time in correspondence and discussion.

Time unfortunately obscures the memory of the source of so much personal knowledge. Moreover, it would be impossible to attempt a complete acknowledgement of all those who have influenced this book. However, no blanket excuse should exclude grateful mention of help from the following:

Berrick Brothers Limited; Binnie and Partners; British Constructional Steelwork Association; Cement and Concrete Association; GAF; Great Ouse River Authority; Institution of Civil Engineers; Francis W. Keyworth, LRIBA; Letraset Limited; W. E. Pegg; Rolls Royce Limited; Rotobord Limited; Sir M. MacDonald and Partners; Scott Wilson, Kirkpatrick & Partners; Soil and Rock Investigation, Limited; Messrs Sketchley (Loughborough); The Butterley Company Limited; The Water Engineer, City of Liverpool; Trent River Authority; and Gunter Wagner of Pelikan Limited.

In addition, extracts from BS4: *1962, Part 1, Specification for Structural Steel Sections: BS 308: 1964, Engineering Drawing Practice;* BS 1192: *1953, Drawing Office Practice for Architects and Builders;* and BS 3429: *1961, Specification for Sizes of Drawing Sheets* are reproduced by permission of the British Standards Institution, 2 Park Street, London, W1 from whom copies of the complete standards may be obtained.

D. V. Jude

Part One
Drawing

Chapter 1
Equipment

Below ground, a soft ground tunnel. This tunnel, 2.54 m in diameter and lined with interlocking precast concrete segments, connects the Metropolitan Water Board reservoirs at Wraysbury and Datchet. The contractor established a world record by completing 313 m of tunnel in a working week. (Photograph supplied by Mitchell Construction, Kinnear Moodie Group)

1. THE MECHANICS OF ENGINEERING DRAWING

Engineering drawings are based upon a pair of axes, x and y, at right angles to each other but not necessarily parallel to the edges of the drawing board. The simplest way of providing these axes is by a tee square and set square on a drawing board, see Fig 1.1.

It is convenient to have the most frequently used, fixed angles on the set squares. This necessitates two set squares, the 'fortyfive', which has two 45° angles as well as the right angle, and the 'sixty thirty' with angles of 60° and 30°.

Well made instruments of good design will last an engineer his lifetime and be a pleasure to use. Some of the cheaper instruments, particularly some plastic set squares and French curves, are badly made, with such rough edges, that they are useless. Feel the working edges of all equipment and reject those that show the slightest sign of roughness; fair wear and tear will roughen the edges soon enough.

1.2 PERSONAL EQUIPMENT

Some of the essential items of personal equipment are listed below:

(a) drawing board
(b) tee square
(c) set squares
(d) adjustable square
(e) protractor
(f) compass and dividers
(g) springbow compass
(h) scales
*
(j) drawing board clips
(k) paper, linen, and film
(l) pencils
(m) erasers.

(*As the letter i, especially as a capital, can be confused with the number 1 or the letter l, particularly in freehand writing, it should not be used as a reference.)

This smaller list includes some items of equipment that you may find useful later on:

(n) mini-board
(o) parallel rule
(p) French curves and flexible rulers
(q) stencils
(r) pens.

Every engineer collects his own tried and trusted bits and pieces. You will soon discover which instruments you prefer and which you can do without. It is inadvisable to buy a large and expensive set of instruments only to discover, with experience, that it contains many items that are rarely if ever used.

1.1 *Drawing board ready for use with paper held in place by clips. The x and y drawing axes are defined by a tee square and set square.*

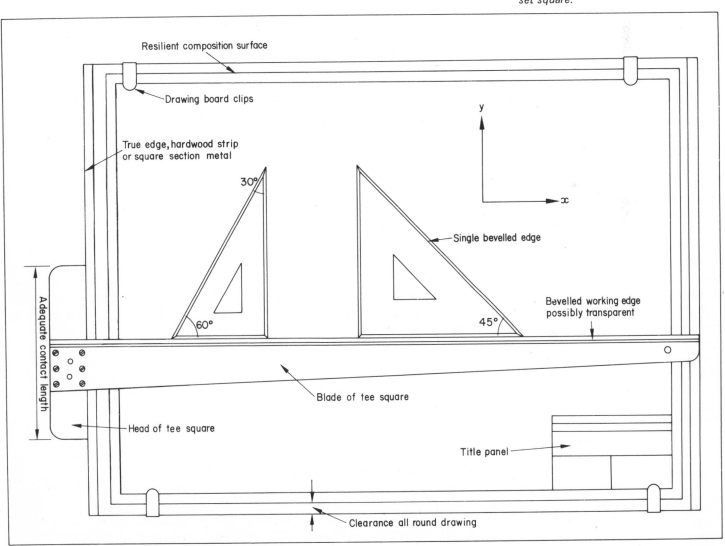

Resilient composition surface
Drawing board clips
True edge, hardwood strip or square section metal
30°
60°
y
x
Single bevelled edge
45°
Bevelled working edge possibly transparent
Adequate contact length
Blade of tee square
Head of tee square
Title panel
Clearance all round drawing

Some of the salient points to look for when buying or using this equipment are listed below.

(a) Drawing boards

Students should have unrestricted use of a drawing board and tee square in their own rooms. If your college or university does not provide or loan equipment, you should buy your own. Ability on the drawing board comes only through practice, and you are unlikely to get enough practice if you have to go to a drawing office whenever you want to draw. It is surprising how much can be accomplished in the odd half hour or less, if the work is already fixed to the board and ready to hand.

The A1 size of drawing, see Chapter 3, is convenient for student use. The drawing boards can be carried about without too much difficulty and the drawings themselves are about the right size for student exercises. The A2 size is probably too small for other than the most elementary exercises. The professional world mostly uses A0 size drawings so you will have little or no use for the A1 drawing board when you have graduated. It is, therefore, unnecessary to waste money on an expensive engineer's drawing board, when a perfectly adequate one can be obtained for half the price.

Plain, wooden surfaced boards need a backing sheet beneath the drawing, an old drawing will do. This protects the surface of the board from over zealous use of the hard pencil; it also prevents a pencil being deflected by the grain of the wood, and results in improved line quality.

Better still, buy a board faced with one of the several resilient compounds from which compass pricks and pencil scars disappear quite quickly.

(b) Tee squares

Tee squares should sit firmly and slide smoothly on the edge of the drawing board, while the blade must lie flat on the paper. To this end, check that:

(a) The head is not too short.
(b) The blade is securely attached to the head, preferably pegged as well as screwed and glued.
(c) The blade is stiff. Some blades, made entirely of transparent material, are rather too flexible.

Tee squares should be hung up when not in use, preferably by the hole in the blade.

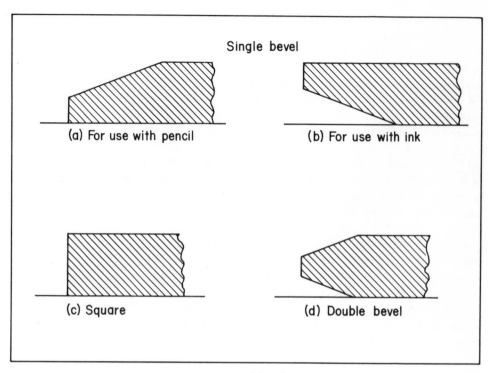

1.2   Types of edge available on set squares.

The double edged tee square, i.e., one with a parallel-sided blade, can be used on either edge of the board and is, therefore, useful to the left-hander who finds it difficult to work right-handed.

Since opposite edges of the board may not be parallel, a drawing must be made with the tee square in the left-handed or right-handed position throughout.

(c) Set squares

As the name implies, these are squares with set or fixed angles. A pair of set squares should be purchased, a 'fortyfive' and a 'sixty thirty', see Fig. 1.1.

A reasonable size is a set square which measures 250 to 300 mm on the longer side adjacent to the right angle. Squares much smaller than this will lead to inaccurate work, while larger ones can be a nuisance to carry about. You may find it convenient to add a much larger 'sixty thirty' set square to your armoury later on.

Of the three types of edge available on set squares, see Fig. 1.2, the single bevel is the most useful. With the bevel uppermost, accurate pencil work can be carried out, while ink can be used with little fear of blots and smudges when the bevel is underneath. The square edge is difficult to use with ink, while the double bevel can be less accurate with pencil and less safe with ink.

A useful addition to any set square, for drawing a line perpendicular to a given line (particularly when the work is not attached to the drawing board), is a line scribed on one face of the set square perpendicular to the hypotenuse, see Fig. 1.3 (a). With the scribed line coincident with the given line, the hypotenuse of the square will lie at right angles to it. This is quicker and more accurate than the usual method of drawing lines at right angles to each other, using two set squares, as shown in Fig. 1.3 (b).

Always check the angles of set squares. Right angles can be checked, as shown in Fig. 1.3 (c), by swinging a set square from the dotted to the solid line position, on a tee square, about the point C.

Devise constructions and checks for the other angles of set squares. It is surprising how often errors are found.

(d) Adjustable squares

The adjustable square is a 'fortyfive' square with a hinged portion so that one of the 45° angles can be opened out to 90° and clamped in any position, see Fig. 1.4.

It is inadvisable to try and make the adjustable square do the work of the two set squares. Students often try this but find it inconvenient and slow.

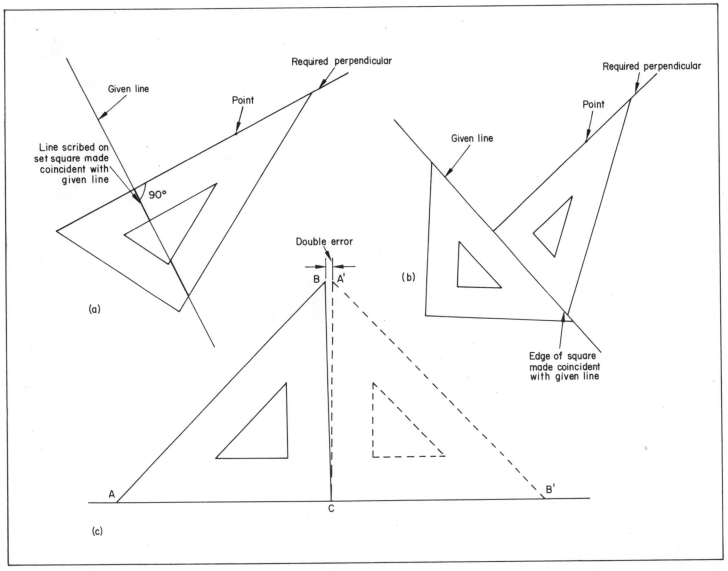

Given line

Required perpendicular

Point

Line scribed on set square made coincident with given line

90°

(a)

Required perpendicular

Given line

Point

(b)

Edge of square made coincident with given line

Double error

B A'

A

C

B'

(c)

1.3 (a) Using a line scribed on the face of a set square perpendicular to the hypotenuse, in order to draw lines at right angles.

(b) Drawing lines at right angles with two set squares.

(c) Checking the right angle of a set square.

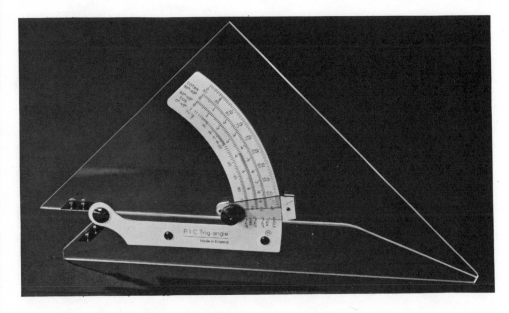

1.4 Adjustable square.

(British Thornton Ltd.)

The main use of the adjustable square is for the quick and accurate construction of parallel lines not on the x and y axes of the drawing, either in normal drawing or for graphical solutions to statistical problems. The protractor is better for setting out given angles.

You will soon find a host of jobs that the adjustable can do quickly and accurately.

(e) Protractor

Although the familiar semicircular protractor measuring about 100 mm across can be used on the drawing board, greater accuracy is obtained from a larger instrument, preferably a whole circle. A good size for plotting chain surveys is about 200 mm in diameter.

(f) Compass and dividers

The type of compass used in school can still be useful to the engineer, particularly with coloured pencils which can be changed so much more quickly than coloured leads in an engineer's compass.

A compass must be capable of drawing an accurate, firm circle in pencil or ink. Probably the best instrument for this is the small beam compass, see Fig. 1.5 (a). The traditional 125 mm compass, shown in Fig. 1.5 (b), is satisfactory for ink work where little pressure is required on the pen to draw a firm line. But, almost without exception, these compasses spring slightly when under the greater pressure required to draw a firm pencil line and thus leave another line the second time round. This fault is overcome by the giant bow with extension arms, shown in Fig. 1.5 (c), but the screw adjustment, although precise, is slow.

The magnificent, but expensive, large bow by *Rotring*, see Fig. 1.5 (d), has fine adjustment by means of a worm wheel held on the curved rack by a spring which allows the wheel to be released for quick coarse adjustment. A pencil attachment and extension arm are included in the set.

Dividers of the 125 mm size, see Fig. 1.5 (e), are invaluable for transferring dimensions from one part of a drawing to another or for dividing a curved or straight line into a number of equal parts, by trial and error. Dividers are *not* used for transferring dimensions from a hand scale to drawings, this should be done directly using a sharp, hard pencil or pricker.

(g) Springbow compass

The precision and ease of adjustment required for drawing small circles is provided

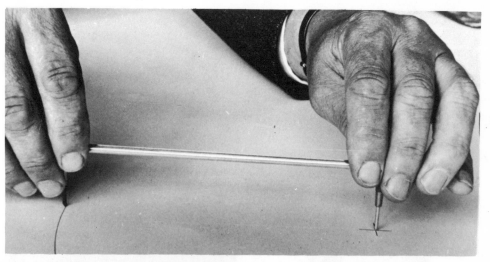

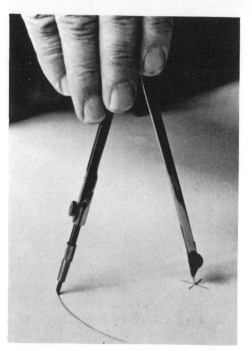

1.5   Typical drawing instruments.
  (a) Small beam compass.
              (British Thornton Ltd.)

  (b) 125 mm compass.
              (British Thornton Ltd.)

  (c) Giant bow.
              (Jakar International Ltd.)

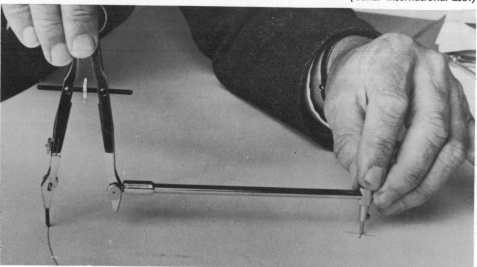

(d) Large bow.   *(Rotring)*

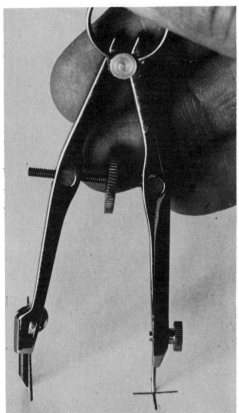

(f) Springbow compasses.
(British Thornton Ltd.)

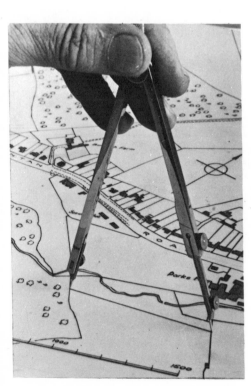

(e) 125 mm dividers.
(British Thornton Ltd.)

by springbows, see Fig. 1.5 (f). The better ones have centre screw adjustment, i.e., the knurled adjusting wheel is between the legs of the bow which makes it possible to adjust the bow with one hand, using the second finger, to operate the wheel, while holding the instrument between the thumb and forefinger.

(h) Scales

The term scale, in drawing, means two things to an engineer, it can be the scale ratio of a drawing, see Chapter 3, or the instrument he uses to measure distances on the drawing. The latter is called a scale because it is marked with the scale ratio of the particular drawing. A ruler, generally speaking, has a more robust edge for ruling lines, while an engineer's scale has a thin edge to reduce parallax errors when pricking off distances. It should not be used for ruling lines, slitting paper, etc.

Always use the correct measuring scale when marking or scaling from a drawing. Never try to get by with a ruler marked in millimetres and convert all dimensions to millimetres on the drawing. Apart from being slow, this will lead to random errors which are hard to find.

Owing to the wide range of scale ratios used by the civil engineer, see section 3.4, his stock of measuring scales could become formidable, unless the more frequently used ratios are combined on one or two scales. The metric BA 7, which contains the ratios:

1 : 50, 100, 200, 2 500
1 : 5, 1, 20, 1 250

or its British Standard equivalent are suitable.

Scales are made 150 mm, 300 mm, or 500 mm long and of several materials and several sections, see Fig. 1.6 (a). A scale should be easy to use, be made of a stable material that will not change its length, have thin edges for accuracy, and be clearly marked.

Flat and triangular section scales do not rock on the drawing board and are perhaps the easiest to use. The flat one has only two edges for graduations, while the triangular one has six but is much more expensive. The oval section provides four edges for graduations but must be held securely at an angle to the paper. This is not difficult to do.

A fully divided scale is subdivided into its smallest units for its whole length; an open divided scale is only subdivided into its

smallest units at one end, see Fig. 1.6 (b).
The fully divided scale is quicker to use so
long as its subdivisions are not too fine and
therefore confusing.

(j) Drawing board clips

The sheet of drawing paper can be held on
the drawing board equally well by drawing
pins, clips, or draughting tape (a special
sticky tape that peels off without damaging
the surface of the paper). Drawing pins are
not generally used as they damage the
corners of the board in time and also rip out
the corners of the drawing when the paper
shrinks. Tape is essential for small drawings
in the centre of the board, since no other fixing can be used. The drawing board clip is
the most convenient way of fixing paper to
the board. The clip cannot damage either the
board or the paper. Drawings are easily adjusted, but, at the same time, they are very
firmly held.

(k) Paper, linen, and film

A reasonable quality cartridge paper will
meet most student needs. It should have a
surface that is smooth enough to take ink
and strong enough to resist rubbing out.

For more complicated work where there is
more chance of errors and rubbing out, a
more expensive, tougher paper should be
used. Some handmade papers have a rough
art surface which is useless for ink however
good it may be for watercolour paintings.

Detail paper is a thin, but strong, semi-
transparent paper with a tough, smooth surface used professionally for sketching out
schemes. It is opaque enough for work to be
seen reasonably well without the white backing sheet that is necessary with tracing paper,
but it is also transparent enough for most
tracing. Photo prints can be made from ink
drawings and clear, dense pencil drawings
made on it. For these reasons, detail paper is
worthy of greater use.

Tracing paper provides a tough, very
smooth working surface and will withstand
plenty of rubbing out. It is more transparent
than detail paper. Good photo prints can be
taken from tracing paper, but pencil work
must be dense for good results. Tracing paper
can be used for normal drawing but requires
an opaque, white backing sheet to make the
work visible, both on the board and in the
hand.

The best dyeline prints, see Chapter 9, are
made from ink tracings or drawings on linen

1.6   *Scales.*
(a) Flat, oval, and triangular section.

*(b) Fully divided and open divided
scales showing similar ratios: metric
1:100 and imperial ⅛ in. to 1 ft (1:
96).   (Faber Castell, W. & G.)*

or polyester film. The former, which is rapidly disappearing because of the superiority of film, is a fine linen impregnated and coated to take ink. It has produced superlative work. Damp affects the dimensional stability of linen. Polyester film is tougher and unaffected by damp or by termites which thrive on linen.

(l) Pencils

Pencils are available in a confusingly wide range of hardness. Hard pencils are denoted by the letter H, while soft ones are labelled B. The scribbling pencil comes from the middle of the range and is the HB. The F pencil in some manufacturers' lists is almost an HB.

Basically, the engineer uses hard pencils, never less than H, usually 3H or 4H, which will take a fine point and will not wear out too quickly, for his detailed work. Soft pencils are used by artists for less detailed work, shading, and blending. Work from soft pencils smudges very easily. They are useless on a drawing board where tee squares and set squares consantly rub over the work. Similarly classified pencils from different makers can vary slightly in their hardness.

For most student work, pencils in the range H to 3H should suffice. A sharp legible line should be produced that will not smudge, but it must be possible to rub it out without going through the paper.

If you choose to use a traditional wooden pencil, you should leave the hardness mark on the pencil and sharpen the other end.

Lead holders or clutch pencils, see Fig. 1.7, are more widely used than the ordinary pencil. In these, leads of different hardness can be used in the same holder. Moreover, sharpening is quicker. Choose one that is neither too heavy nor too light and that will not slip out of sweaty fingers.

Engineers use both conical and chisel pointed pencils, as shown in Fig. 1.7 (b) and (c). Both points start with 5 to 10 mm of exposed lead, see Fig. 1.7 (a). Use a draughtsman's block of fine sandpaper to prepare either point and to keep it sharp in use, see Fig. 1.8. If the conical point is kept sharp and slender, it will suffice for most drawing. It can be kept reasonably sharp in use by rotating the pencil as the line is drawn. If many long straight lines are to be drawn, a chisel point should be used as it keeps sharp longer.

Compass pencil points are mostly elliptical, see Fig. 1.7 (d). This is achieved by rubbing

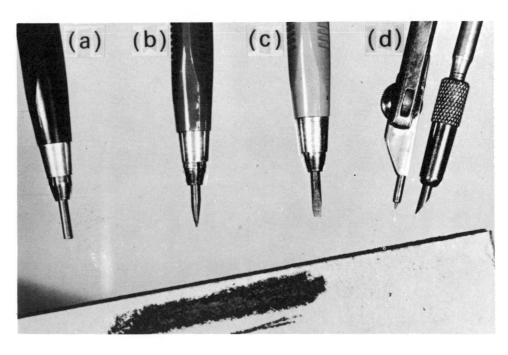

one face only on the sandpaper block. Such a pencil point can be located close to the compass point so that small circles can be drawn. An elliptical point wears almost as well as a chisel point. Moreover, this is a conveniently quick way of pointing any pencil.

(m) Erasers

Pencil lines can be removed by almost any one of a wide range of rubbers, from ones actually made of rubber through art gum blocks to the newer transparent erasers that do not create quantities of crumb-like debris when used.

Ink work has to be removed with something harsher, either an ink rubber or glass fibre brush. The latter is very effective on most surfaces and very quick, but it can damage the working surface unless the surface is very strong. The little pieces of glass fibre that wear off in use are very irritating to the skin. Many draughtsmen do not use these erasers for this reason alone.

If selective erasing has to be done, an erasing shield of metal or plastic, see Fig. 1.9, can be used. The work to be erased is framed in one of the many apertures. Also shown in Fig. 1.9 is a type or ink eraser in which the core of eraser, about 5 mm in diameter, is sheathed in wood, the same size and shape as a wooden pencil. These erasers are sharpened like a pencil to expose a sharpish point of eraser that can be quite selective in use.

The following items of personal equipment

1.7 Clutch pencils, showing types of point, and a sandpaper block.
(a) Exposed lead.
(b) Conical point.
(c) Chisel point.
(d) Elliptical point on a compass lead.

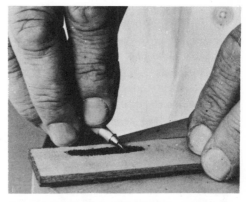

1.8 Using a sandpaper block.

1.9 Erasing shield with type or ink eraser.

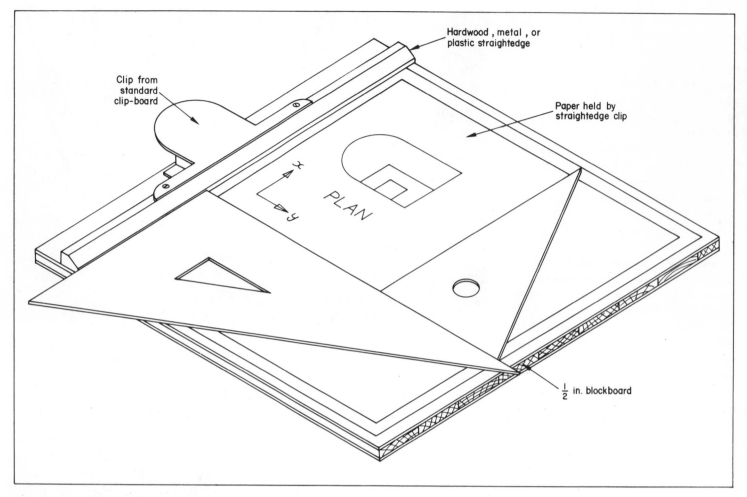

Clip from standard clip-board

Hardwood , metal , or plastic straightedge

Paper held by straightedge clip

PLAN

$\frac{1}{2}$ in. blockboard

are not essential for beginners, but you may soon find them useful.

(n) Mini-boards

Small, accurate drawings are often required for design notes and laboratory reports. While these can be done perfectly well on the A1 board, it is often more convenient and certainly less tiring if a smaller board is available that can be used when sitting at a desk or table.

A simple, small board with a tee square would suffice, but the device shown in Fig. 1.10, which can be made quite cheaply, has proved convenient and quick to use. The paper is held along one edge by a sprung graduated straightedge along which a set square can slide to provide the x axis of the drawing, while the y axis is determined by another set square. Commercial versions of this board are available.

One approach to the smaller, more accurate, engineering drawings discussed in section 3.2 is the *Rotobord* shown in Fig. 1.11. A single straightedge is provided while the drawing, up to A3 size, is mounted on a circular rotat-

able board. The straightedge moves on a vertical axis, like a traditional tee square, but with a precise slide with optional click stops to facilitate measurements or precise shading. The rotating board also has click stops for the usual 30°, 45°, 60°, and 90° angles plus a precise protractor scale and a clamp for any angle. It is convenient and very quick to use.

(o) Parallel rule

Small, accurate drawings can also be prepared by using a parallel rule in conjunction with a set square which has a line scribed perpendicularly to its hypotenuse, as described in section 1.2 (c). Parallel lines can be drawn by using two set squares, but the parallel rule ensures greater accuracy. Guide lines for printing and tables for the results of experiments can also be drawn quickly with the aid of a parallel rule.

(p) French curves and flexible rulers

Two fundamental faults ruin a drawing, bad printing and badly drawn irregular curves. Smooth irregular curves can only be drawn with the aid of a French curve or a flexible

1.10 Homemade mini-board for desk use to make accurate A4 drawings for design notes and laboratory reports.

ruler.

A set of French curves consists of several shapes in transparent plastic, see Fig. 1.12 (a). They are made up of smooth but irregular curves, bits of parablolae, bits of ellipses, etc. In use, a French curve is chosen and moved about over the line to be drawn until the part that best fits the curve is found. Your first dozen or so attempts may well result in smooth curves joined by sharp corners. However, constant practice and considerable overlapping of the French curve on the line already drawn will result in smoother joins. Do not try to draw too much at once. Always make sure that the French curve is tangential to the part of the line already drawn.

Some workers prefer the flexible ruler, see Fig. 1.12 (b), which has a flexible outer case on a malleable core which will bend but retain its bent shape. The core can become brittle after some use and snap, while previous curves cannot always be completely removed.

(q) Stencils

There is no substitute for neat, stylish, free-hand printing, but few of us can achieve this, especially in titles where larger letters are used. You may, therefore, be advised to use letter stencils in the early days for the larger letters, if only to learn a style and overcome the depressing effect of seemingly puerile lettering on an otherwise acceptable drawing.

Stencils can be used with pencil or ink.

They should be dispensed with as soon as possible, if only because they are so slow to use.

(r) Pens

You should become familiar with the use of pens for drawing as soon as possible.

The traditional ruling pen, see Fig. 1.13 (a), is the cheapest and most versatile device for use with ink. It can produce unsurpassable work in skilled hands. However, students sometimes find that they do not have the time or the patience to become skilled in its use. This type of pen can become clogged fairly often in use as the ink dries. After cleaning, it is difficult, even with a calibrated adjusting screw, to match exactly the thickness of the lines previously used.

Figure 1.13 (b) shows the *Pelikan Technos* pen with its relatively cheap interchangeable nibs. Three types of nib are available, each in a range of different sizes, so that exact line thicknesses can be reproduced at will. The type B nib has two leaves, like the traditional ruling pen, and will draw lines with sharper ends than the type D which is also for drawing lines. The type C has a slightly rounded tip so that it is smoother in action for printing if it is not held absolutely normal to the paper. The cartridge ink supply in the body of the pen is adequate for weeks of continuous work, while an unblocking device is designed to restart the flow should the ink clog when the pen is not being used.

Figure 1.13 (c) shows the forerunner of the *Technos* pen, a typical stylus drawing pen with interchangeable nibs. These are more expensive or messy to change than the *Technos*, depending on which part you choose to change. The refillable reservoir gives weeks of use. A fine wire in the hollow stylus both meters the supply of ink and can be shaken up and down to unclog dried ink. Nibs with rounded tips are available for smoother printing, but not twin leafed nibs for sharper ended

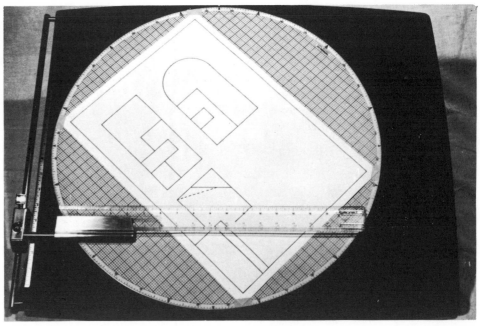

1.11 Rotobord *for precise drawings up to A3 size.*

1.12 *Drawing irregular curved lines.*
    (a) French curves.   *(Jakar)*
    (b) Flexible ruler.   *(Jakar-flex)*

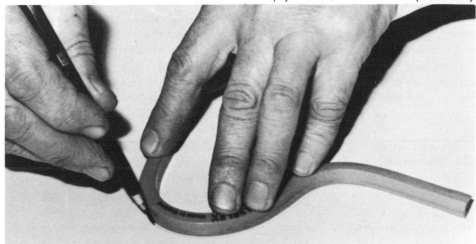

lines. Development of this kind of pen is going on all the time.

1.3 DRAWING OFFICE EQUIPMENT

The list below contains some of the more important items of equipment that you will come across in a drawing office. They are all designed to speed or improve draughting:

  (a) draughting machine
  (b) powered erasers
  (c) railway curves
  (d) large beam compasses
  (e) straightedge.

(a) Draughting machine

Instead of a drawing board used on a bench with a tee square, you may find yourself faced with a draughting machine. Fundamentally, this consists of a drawing board supported on a stand which gives a wide range of working positions by varying both the height and the inclination, see Fig. 1.14.

The axes of the drawing can be provided by:

  (a) A parallel motion straight edge and set squares, Fig. 1.14 (a).
  (b) A pantagraph arm or other system with drawing head, Fig. 1.14 (b). The term draughting machine is frequently reserved for this type.

The parallel motion is by far the more popular piece of equipment. It provides a sturdy straightedge, controlled by wires or chains, that moves smoothly up and down the board, always remaining parallel to the x axis of the drawing. Set squares and adjustable squares are used in exactly the same way as with the tee square. Since far less effort is required to keep the straightedge firmly in place, there is an increase in drawing speed and a reduction in fatigue.

The pantagraph arm with drawing head (or draughting machine) replaces set squares as well as the tee square. It consists of a drawing head mounted on a linkage which keeps it parallel to the axes of the drawing. The drawing head has two scales set at right angles to determine the x and y axes. The scales can be reversed or replaced by others so that the draughtsman always has his scales ready to hand. The head will rotate to any angle where it can then be clamped. Thus angles can be set out or, more conveniently, parallel lines can be drawn very quickly. The drawing head has click stops usually at 15° intervals so that the common angles, 30°, 45°, and 60°, can be

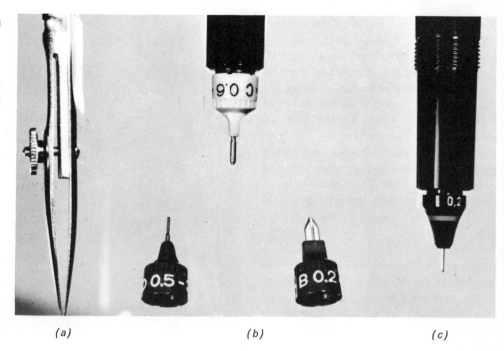

(a)   (b)   (c)

1.13 Drawing pens.
   (a) Traditional ruling pen.
   (b) Pelikan Technos with three interchangeable nibs.
   (c) Typical stylus pen.   (Rotring)

selected and repeated exactly without fiddling with the protractor scale.

Diagonal lines between widely spaced points cannot be drawn on a draughting machine without a long, separate straightedge.

(b) Powered erasers

The powered eraser is rather like an electric drill only less powerful and far less noisy. Hard ink rubbers, about the size of the rubbers used on the ends of scribbling pencils, are fitted into a chuck which revolves at about 8000 rpm. They soon remove ink lines, not to mention the paper or film, if not used with care.

(c) Railway curves

For roads and, as the name implies, for railways, very flat curves often have to be drawn, the centres of which are sometimes outside the drawing office, let alone off the drawing board. Bridge and dam profiles too require flat curves, i.e., arcs of large radius. These are provided on a set of templates each one an arc of different radius.

(d) Large beam compass

The small beam compass mentioned in section 1.2 (f) is limited to a radius of 200 to 250 mm. Drawing offices normally possess a larger instrument with a radius of about a metre. The beam may be in sections for convenience of storing or for carrying about. The point of the pen or pencil should have a screw device for fine adjustment of the radius. Beam compasses are used quite extensively in plotting surveys, see section 8.5.

(e) Straightedge

A straightedge is needed to draw long diagonals when using a draughting machine, section 1.3 (a), or in plotting surveys, section 8.5 (b). Straightedges are of steel or stable plastic and should be safely stored away when not in use, either hung up or in a box. A straightedge should never be used for slitting drawing paper or for trimming the edges of photo prints with a razor blade, but, alas, this is so often its fate.

1.4 FIRST STEPS IN DRAWING

Each step described below is simple and obvious, but observance of the sequence will speed the preparatory work for a drawing.

Attach the sheet of paper to the board by clips with the top edge parallel to the x axis defined by the tee square. Use a backing sheet if the board is not faced with a resilient composition.

Check that the paper is the right way up. The wrong side often has a pronounced grid or waffle pattern from the manufacturing process.

If the paper is carried about rolled up in a cardboard or plastic cylinder, it is better to roll it working face outwards, not inwards which is the natural way to do it, so that the induced curl of the paper is easier to control when it is on the board.

Draw a border 10 to 15 mm from the edge all round the sheet and insert a title panel, see Fig. 3.1 and Fig. 3.2. For normal class exercises, the title panel can be simplified, but it must be there.

Now decide where the work is to be fitted into the sheet of paper. If there is plenty of room, make sure that the work is symmetrically spaced over the whole sheet and not squashed into one corner. At this stage, even experienced draughtsmen find it worthwhile sketching the layout, sometimes roughly to scale, on a piece of scrap paper. Decide on the scale of the drawing and see whether all of the scheme or exercise can be shown on one sheet or whether it should be spread on to others for better resolution of detail.

Once the sketch layout has been satisfactorily completed, lay the work out lightly, by drawing the centre lines and block outlines on the drawing sheet, to make sure that it fits in neatly and symmetrically. With the block outlines in place, the overall detail can be worked up. If an orthographic projection or a similarly related set of views is being drawn, complete all the views lightly before lining in firmly. There is a great temptation to line in each bit as you complete it, but resist this, it is so much easier to rub out draft lines than the final heavy ones. Moreover, there is less danger of smudged lines.

Curves and radius fillets should be drawn in first as it is easier to fit tangents to curves than *vice versa*. Make all square corners sharp and obvious. Lines should start crisply at full strength, width, and density.

The most difficult part of a drawing begins when all the line work is complete, i.e., the finishing which consists of dimensioning, shading of sections, and lettering. Many otherwise reasonable drawings are ruined at this stage. All of these operations should be carried out simultaneously, for example, a hole may have to be left in a block of sectioning for a dimension or note, lead lines for notes may have to be moved to prevent confusion, or perhaps dimensions may have to be inserted somewhere else for the sake of clarity.

Engineering drawing is in some ways like a competitive sport. You only become good at it by dint of continuous practice. Graduation from barely adequate to good is imper-

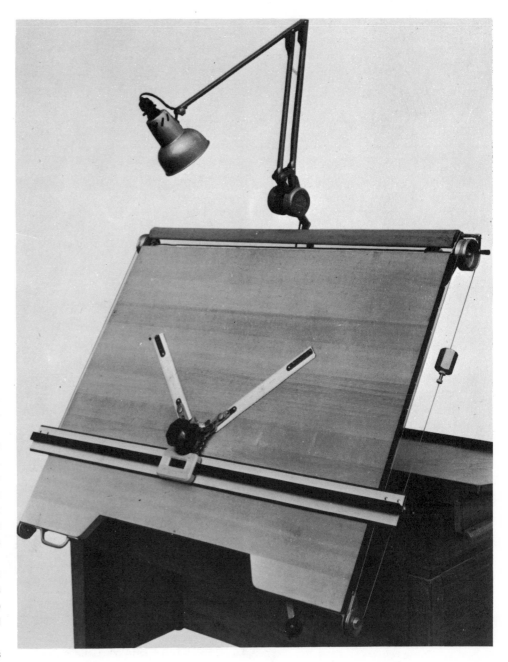

ceptible. An athlete does not suddenly begin to win races nor does the dinghy helmsman suddenly find himself among the leaders. Similarly, the student cannot expect instant success on the drawing board. Unfortunately draughtsmen, like athletes, can lose prowess without practice. The graduate should not expect to be able to compete directly with a good, experienced draughtsman, but he must be able to express his thoughts clearly, neatly, and economically.

1.14 *Draughting machines.*
  (a) *Board on adjustable stand with parallel motion straightedge. (Also equipped with a draughting head.)*
  (Admel)

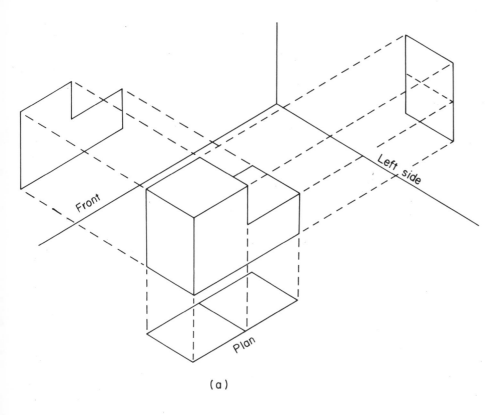

(a)

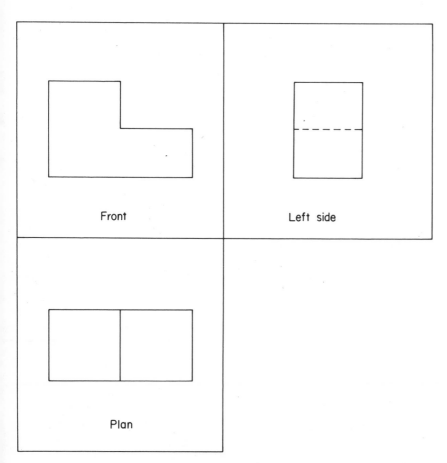

(b)

2.7 *First angle projection.*
  (a) How the orthographic views are cast forward to the planes of projection.
  (b) The resulting conventional arrangement when the cube is laid out flat.

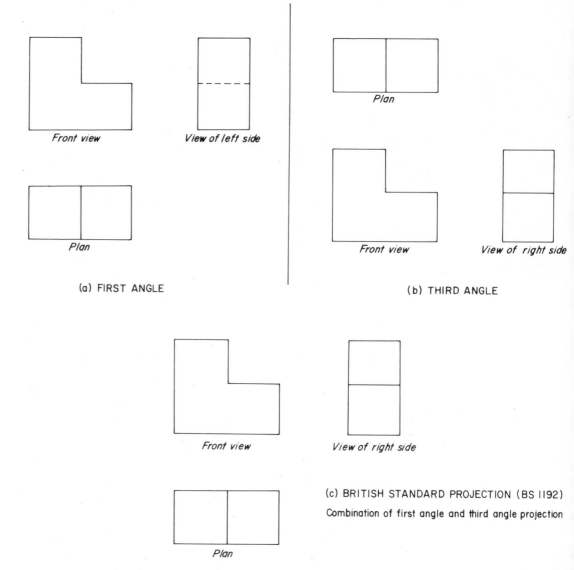

2.8   *Comparison of projections.*

In relation to the elevation, end views are placed so that they are in third angle projection and plan views in first angle projection.

For building drawings this method has been practised very widely and combines the advantage of both first and third angle projection.

It is recommended that all elevations should be properly identified, either with descriptive notes, or joined with projection lines or by some other suitable means.

2.5 APPLICATION OF ORTHOGRAPHIC PROJECTION

All rules are made to be broken so long as the discipline behind them is understood.

In Fig. 2.9 the general outline of a simple structure is shown, with the three views laid out strictly in accordance with the rules of

third angle projection. The views themselves, however, have been modified in order to show structural detail more clearly.

The plan is quite straightforward, there is very little hidden detail which has all been shown.

A longitudinal section on the centre line is shown in place of a normal front view. As this particular structure is buried in the ground, a normal front view would be seen through the ground so that the entire view would consist of hidden detail and give rise to a confusing mass of dotted lines.

Similarly, a normal side view would show too much hidden detail, most of which would be buried in shading showing earth in section, therefore a sectional view has been used instead.

Owing to the shape of the flume and the profile of the water as it flows through it, the

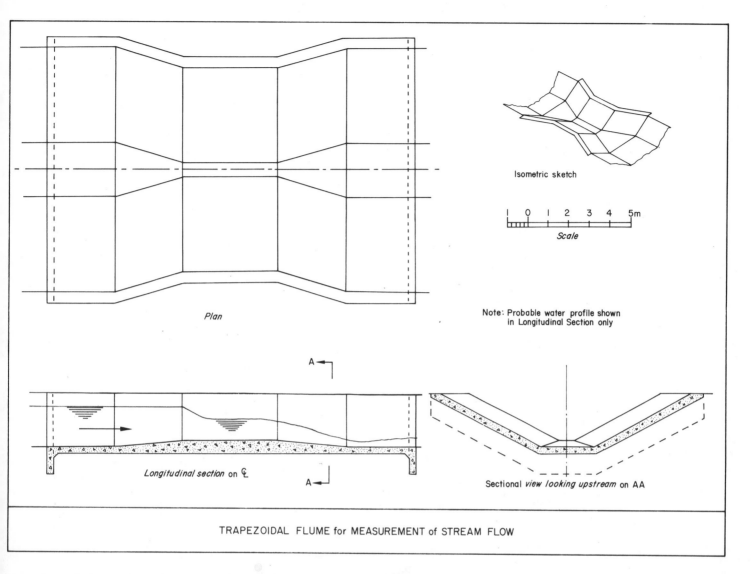

Isometric sketch

Scale

Note: Probable water profile shown
in Longitudinal Section only

Plan

A

Longitudinal section on ℄

A

Sectional *view looking upstream* on AA

TRAPEZOIDAL FLUME for MEASUREMENT of STREAM FLOW

plan view of the water surface would be complicated therefore it has been omitted, but a note stating this has been added. This expedient has also saved the time that would otherwise have been spent in preparing the curves of intersection.

Most civil engineering schemes include structures which are too large to be shown by the conventional application of third angle projection, therefore the rules have to be broken. Figure 2.10 shows the outline of an earth dam, fundamentally a very simple structure to draw, but too large for the side views (represented by cross sections) to be shown alongside the front view. The required cross section is shown below but is carefully referenced.

An earth dam is quite simple and its detail can sometimes be shown to a scale small enough for it all to be contained on one drawing. More complicated structures will have to

be shown on a larger scale to resolve their detail and thus will spread to several drawings.

However useless it may be for dimensions and small detail, a small drawing of a large project, giving several related views on one sheet, can be very useful for showing the whole scheme to lay observers. This type of drawing will probably be valued by senior engineers who have not the time to digest several drawings but who, nevertheless, must know the overall pattern of the scheme.

2.6 SECTIONS

It is frequently necessary to show a structure in section, i.e., cut open, either to help with the general presentation, as with the flume in Fig. 2.9, or to show internal detail, as with the earth dam in Fig. 2.10.

The symbolic shading of sections is described in Chapter 3, but must be applied with discretion and should not be allowed to over-

2.9 *Suitable method of using third angle projection to show a small and simple structure.*

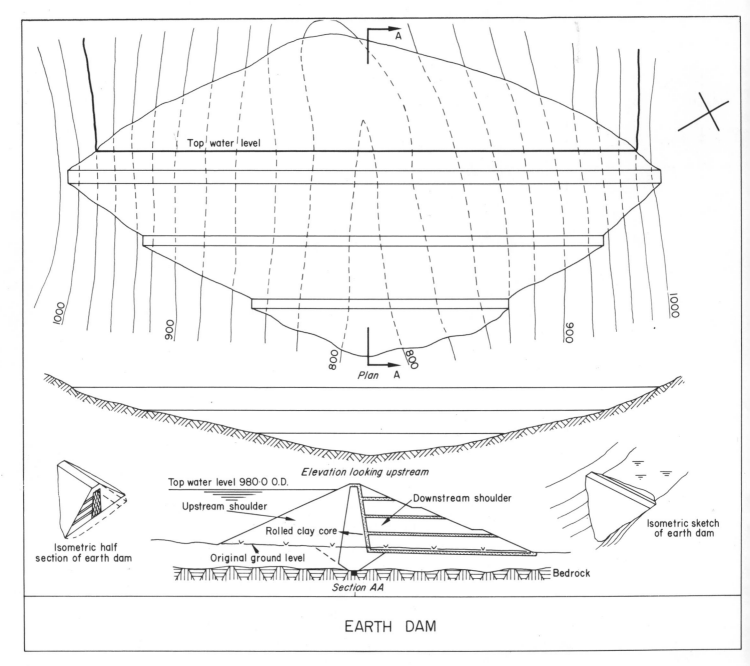

EARTH DAM

2.10 *Suitable arrangement of orthographic views of a large structure.*

power other detail. For example, with the concrete retaining wall in Fig. 2.11, where a relatively large expanse of sectioned concrete must be shown, the whole effect is lightened by shading selected patches and leaving quite large areas clear. Such techniques may be used, so long as the required effect is obtained and there is no possibility of error. Areas should be left free for dimensions, some hidden detail, or other symbols.

While the symbolic shading immediately identifies a section, the location of a section can only be shown by references in other views. Unless a section is a particularly awkward one, it is seldom necessary for the lines

of the cut to be superimposed on the detail of the view in which the location of the section is shown. The location can usually be shown by short, heavy reference lines outside the general detail, as shown in Fig. 2.10. Arrows should show the direction from which the section is viewed.

Sometimes one section can do the work of two or perhaps more by changing the line of cut at suitable points. If a section does wander in this way, its precise line must be shown in another view, see Fig. 2.12.

The layout of any drawing must be carefully planned in advance. This is particularly important when choosing sections. They must

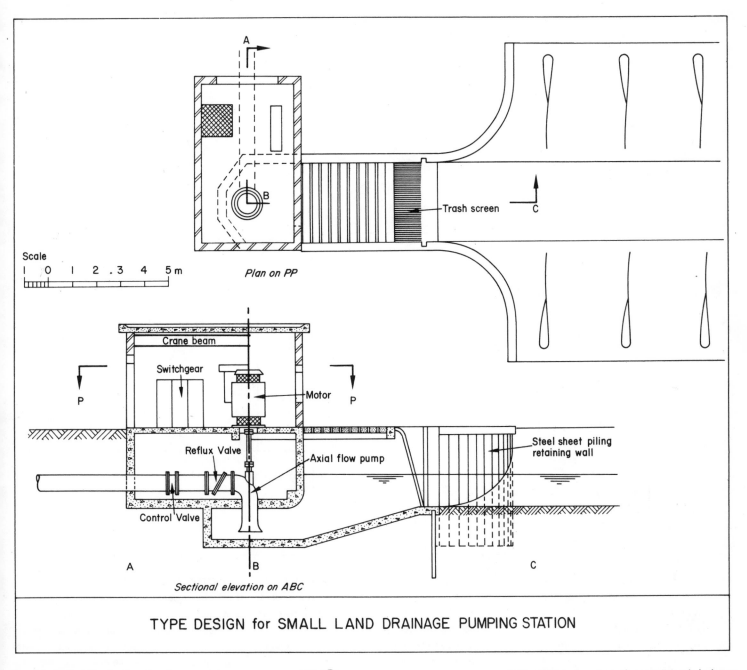

Scale
1 0 1 2 . 3 4 5 m

Plan on PP

Trash screen

Crane beam

Switchgear

Motor

Reflux Valve

Axial flow pump

Control Valve

Steel sheet piling
retaining wall

A   B   C

Sectional elevation on ABC

TYPE DESIGN for SMALL LAND DRAINAGE PUMPING STATION

be selected early in the drawing process so that reference lines can be shown before other detail is added to avoid possible confusion.

2.7 AUXILIARY VIEWS

It sometimes happens that because of the irregularity of its outline, the true shape of one or more faces of an object cannot be shown in any of the normal orthographic views. This will occur when the face to be shown is not parallel to one of the axes chosen for projective purposes. The solution is to project a view from the face in question on lines normal to the face into a suitable space on the drawing, as shown in Fig. 2.13 (a).

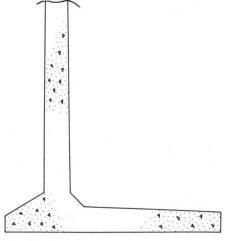

2.12 *Plan and secton of a small land drainage pumping station. Note how the section is taken on the line ABC in the plan.*

2.11 *Section of a reinforced concrete retaining wall with conventional shading limited in extent for lighter effect.*

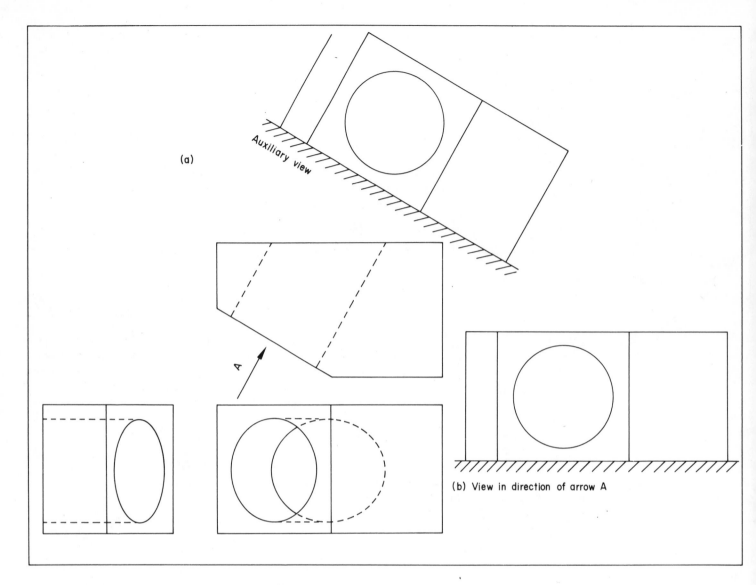

(a)

Auxiliary view

A

(b) View in direction of arrow A

2.13 *The orthographic view of a face, not on the axes of projection, projected on to an auxiliary plane.*
  (a) Conventionally.
  (b) In a more practical manner.

If a large number of auxiliary views seem to be necessary, rearrangement of the object within the planes of projection may help to reduce the number.

Although civil engineers should be capable of projecting auxiliary views as shown in Fig. 2.13 (a), they may never use the technique in practice. Once again, the size of the objects to be depicted mitigates against the strict observance of the rules of projection. Whether there is space on a drawing to project a view at an angle or not, it is inconvenient to have views at odd angles on a drawing, especially if the drawing is large and has to be folded for use on a windswept site.

The usual procedure is to indicate, by a heavy arrow on the plan, the line on which the auxiliary view is seen and then to locate the view on the same axes as the rest of the drawing and to label it, for example, 'View in direction of arrow A', see Fig. 2.13 (b).

2.8 PICTORIAL DRAWING

If the object being drawn is at all complicated, or it must be described to a lay observer who has not been trained in orthographic projection, a reversion to three-dimensional illustration is necessary. In some cases a quick, free-hand sketch will be sufficient, but more complicated examples will require some formalized technique if proportion and scale are to be maintained.

True three-dimensional illustration shows perspective. This is dealt with in Chapter 6. There are several simpler and quicker ways of showing depth without the complication of perspective.

(a) Oblique projection

Perhaps the quickest pictorial method is oblique projection, which consists, in essence, of adding bogus depth to an orthographic front view of an object, see Fig. 2.14.

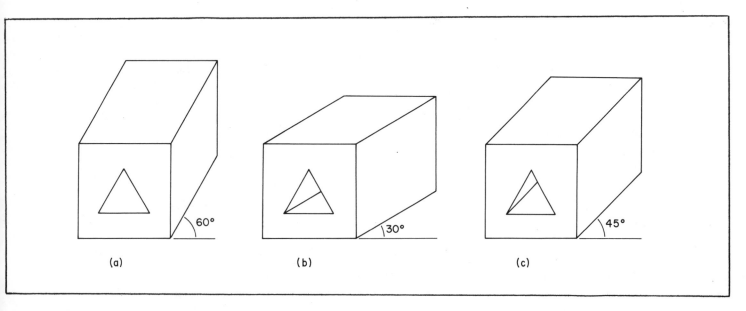

(a)   (b)   (c)

As the top and sides of an oblique projection are badly distorted, it is essential that the orthographic view chosen as the basis for the illustration should contain the details to be clarified.

The receding axis can be at any angle to the front view and can be varied to show more of the top or more of the side, see Fig. 2.14 (a) and (b). Unless a particular face is to be accentuated, the receding axis should be at 45°, see Fig. 2.14 (c).

(b) Axonometric projection (BS 1192: 1953)

Axonometric projection, as shown in Fig. 2.15, is sometimes used by architects.

Like oblique projection, a bogus effect of depth is obtained by adding a third dimension to an orthographic view, the plan in the example shown. If the orthographic plan view is set at an angle to the horizontal axis of the drawing, more than one face of the object can be shown.

Although a remarkably realistic impression is obtained with little complication, the use of true heights makes the building look too tall. A simple halving of the vertical measurements as shown reduces the height too much, three-quarters scale would perhaps be better. A more realistic effect could be obtained by using isometric projection without recourse to differential scales.

(c) Isometric projection

A true three-dimensional effect can only be obtained if all the faces of the object are at an angle to the plane of projection and, therefore, suffer distortion.

Surprisingly, if a cube is put on a table and viewed so that all the faces are at an angle to the line of vision, the infinite number of different views fall into only three categories. These are shown in Fig. 2.16.

In isometric projection, the cube is viewed symmetrically so that all its edges seem to be of the same length, that is they must all be at the same angle to the plane of projection. As the name isometric implies, there is equal measure, all the sides are of the same length.

In dimetric projection, the viewpoint remains on the central axis but is lowered slightly or raised so that the top of the cube is foreshortened or made to look longer. In this way, the edges of the cube appear with two different lengths in the projection, 'two measures' in fact as the name infers.

Trimetric projection results from completely random or asymmetrical viewing of the cube so that the sides on all three axes of the cube appear with different lengths and there are 'three measures'.

Dimetric and trimetric projection are seldom used. If such a complication is warranted, it is

2.14 *Oblique projection where depth is given to an orthographic view by running adjacent surfaces back an arbitrary distance at any suitable angle.*

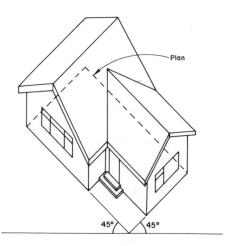

2.15 *Axonometric projection where realistic but arbitrary depth is given to an orthographic plan view (BS 1192).*

2.16 *Comparison of geometric three-dimensional projections.*

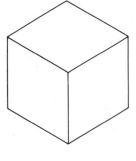
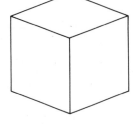
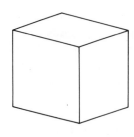

(a) Isometric   (b) Dimetric   (c) Trimetric

probably worthwhile making a perspective drawing as described in Chapter 6.

Figure 2.17 shows, in third angle orthographic projection, a cube set up at an angle to a plane to produce an isometric projection.

The front view and plan show how the opposite corners of the cube, A and G, are on a line normal to the plane of projection. In this position, all the edges of the cube are at an angle θ to the plane of projection.

The side view in Fig. 2.17 is, in fact, the isometric projection. The plan is not. It is dimetric, as the edges make two different angles with a horizontal plane of projection.

In the isometric projection in Fig. 2.17 the three adjacent edges of the cube AB, AE, and AD, representing the three axes of the cube, are equally spaced at 120° round the point A, which is in contact with the plane of projection. This is the basis of all isometric construction. Figure 2.18 shows how the axes can be set up at any angle so long as they are equally spaced round a central point.

It will be seen in the isometric view, which is the right side view in Fig. 2.17, that although the three edges, AB, AE, and AD have the same length, the two diagonals, AC and BD, do not, although they are the same length as each other in the original cube. There must, therefore, be some axes on which measurements can be made, and some axes which suffer distortion and cannot be used for measurement. The principal axes of the cube, represented by the edges AB, AE, and AD, are the only ones that do not distort and are, therefore, the only ones on which measurements can be made. Points on the faces of the cube away from the edges must be located by offset measurements from the edges, see Fig. 2.22.

Isometric scale. If an isometric drawing is being used by itself as a pictorial representation of the object, the actual measurements of the object or true dimensions can be used for plotting along the isometric axes. However, if the isometric view is to be used alongside the orthographic views it is to clarify, the dimensions for the isometric view must be reduced to cos θ times their true value otherwise it will appear very much larger than the orthographic views and will be misleading, see Fig. 2.19.

To determine the value of θ and cos θ refer to Fig. 2.17 and consider the sides of the cube to be x units long. The length of the diagonal

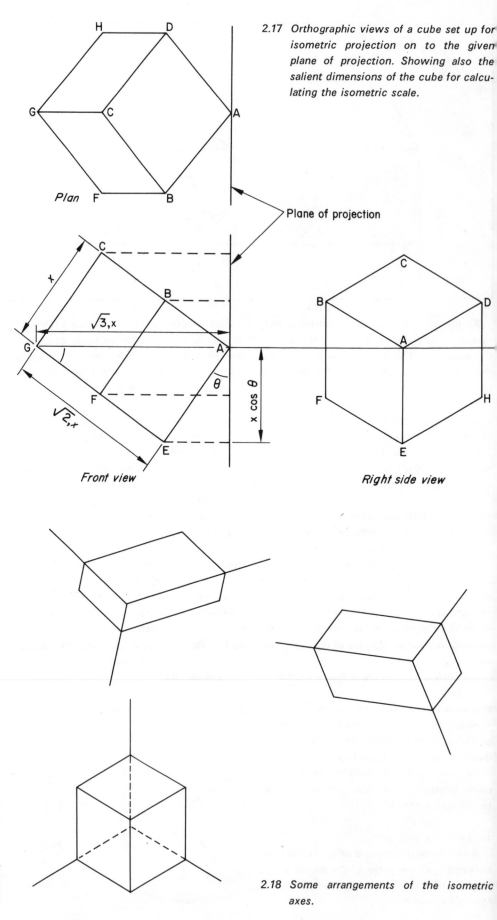

2.17 Orthographic views of a cube set up for isometric projection on to the given plane of projection. Showing also the salient dimensions of the cube for calculating the isometric scale.

2.18 Some arrangements of the isometric axes.

of a side is $\sqrt{2.x}$, while the opposite corners of the cube are $\sqrt{3.x}$ apart (by application of Pythagoras' theorem).

Hence cos θ is

$$\sqrt{\frac{2.x}{3.x}} \quad \text{or} \quad \sqrt{\frac{2}{3}} \quad \text{or } 0.816,$$

or $\theta = 35°16'$.

Thus for a realistic view, all measurements should be reduced to 0.816 times their true value before being used in an isometric projection. This can most easily be done graphically, as shown in Fig. 2.20. Measure the true length along the line inclined at 45° and drop a perpendicular to the line inclined at 30°, which will then contain the isometric length.

Circles in isometric projection. Figure 2.21 is an isometric view of a cube with circles on the three visible faces. The circles appear as ellipses with their major axis parallel to the longer diagonal of the cube face.

As isometric projection is a pictorial impression, an approximate construction of the ellipses is justified.

An ellipse can be approximated to two pairs of circular arcs, see section 5.4 (d). In isometric work, the arcs span the spaces between the isometric axes. In Fig. 2.21 the arcs will be seen to span the spaces between X and Y and also between XX and YY. Here is a quick method of drawing the ellipses:

(a) Set off the diameter along an isometric axis, lines XY.

(b) Draw the enveloping quadrilateral ABCD.

(c) Draw the diagonals of the enveloping quadrilaterals, these will contain the major and minor axes of the ellipse.

(d) Join the corner A of the enveloping quadrilateral to the ends of the diameters X.

(e) The centres P of the smaller arcs are at the intersection of lines AX and line BD.

(f) The larger circular arcs have their centres at the obtuse corners of the enveloping quadrilaterals.

Remember that this is a quick, approximate method. The ellipses so formed deviate from the exact shape in some places and will not always meet straight lines drawn by offset methods as they should. If much of the picture is a combination of curves and straight lines, the whole should be constructed by offsets, see the next section.

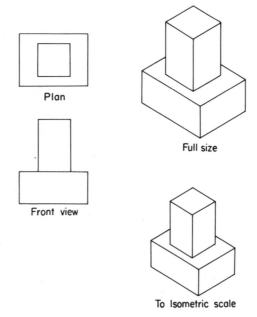

Plan

Front view

Full size

To Isometric scale

2.19 *The greater realism of an isometric view drawn to isometric scale than one drawn full size, in comparison with orthographic views.*

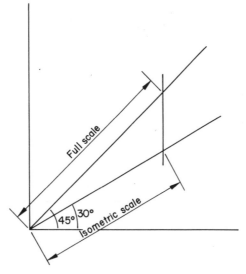

Full scale

45° 30°

Isometric scale

2.20 *Graphical construction of isometric scale.*

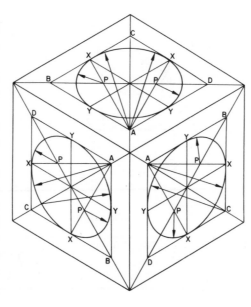

2.21 *Circles in isometric projection. Showing the construction of ellipses on three faces of a cube.*

Irregular shapes in isometric projection. It is convenient to classify any shape, other than a circle, as irregular for isometric projection since the method of construction is the same: to envelope the shape in a square or rectangle orientated to lie on the isometric axes. Measurements can be made along the sides of the enveloping rectangle to locate the points necessary to draw the shape.

In Fig. 2.22 a parabola, drawn in this way, is shown. Points 1, 2, 3, 4, 5, and 6 are located by means of offsets from sides AD, BC and A0, 0B which can be drawn on the isometric axes without distortion.

Figure 2.23 shows two orthographic views of an hexagonal nut. In the plan view, the nut is shown enveloped by a rectangle ABCD which is then drawn on the isometric axes. The corners X of the nut can be located by measurements along the isometric axes. The circular hole is represented by an ellipse with its major axis horizontal.

2.9 INTERPRETING ENGINEERING DRAW-INGS—VISUALIZATION EXERCISES

An engineer reads and interprets many more drawings than he produces himself. His own drawings are the result of a close study of other drawings. If he is well trained in projection, he quickly builds up a useful mental image of a scheme from a strange set of drawings. This is rather different from the slower process of producing orthographic views of his own ideas.

The following exercises, making three-dimensional views from orthographic ones and vice versa, are designed to sharpen your understanding of the conventions of projection laid down in this chapter.

Isometric sketches are quickly and accurately made freehand on isometric paper, i.e., paper printed with axes 120° apart, which is available in blocks. Orthographic or oblique views are made more quickly and neatly on paper with lightly printed 5 mm squares.

References and further reading
British Standard 308: 1964, *Engineering Drawing Practice*

British Standard 1192: 1953, *Drawing Office Practice for Architects and Builders*

Anderson and West: *Engineering Drawing*, Heinemann

French and Svenson: *Mechanical Drawing*, McGraw-Hill Book Company Inc.

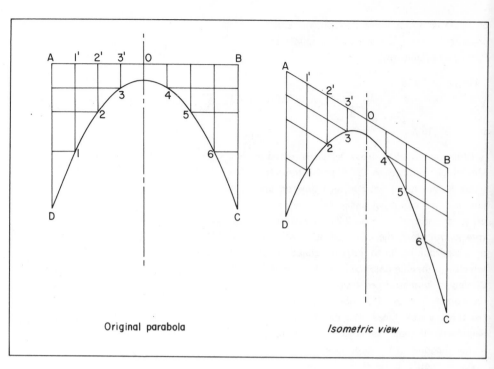

Original parabola   *Isometric view*

2.22 *Construction of irregular curved shapes in isometric views by offsets from the isometric axes, seen applied to a parabola.*

2.23 *Construction of irregular plane shapes in isometric views by offsets from an enveloping rectangle that can be constructed on the isometric axes.*

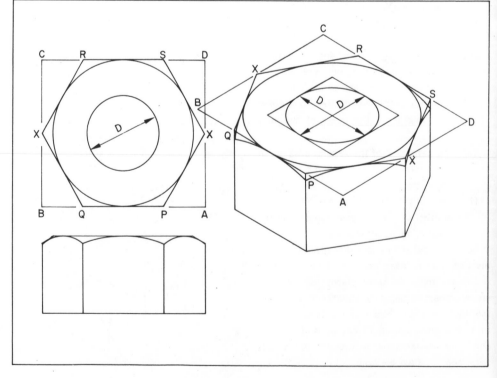

2.10 EXERCISES.

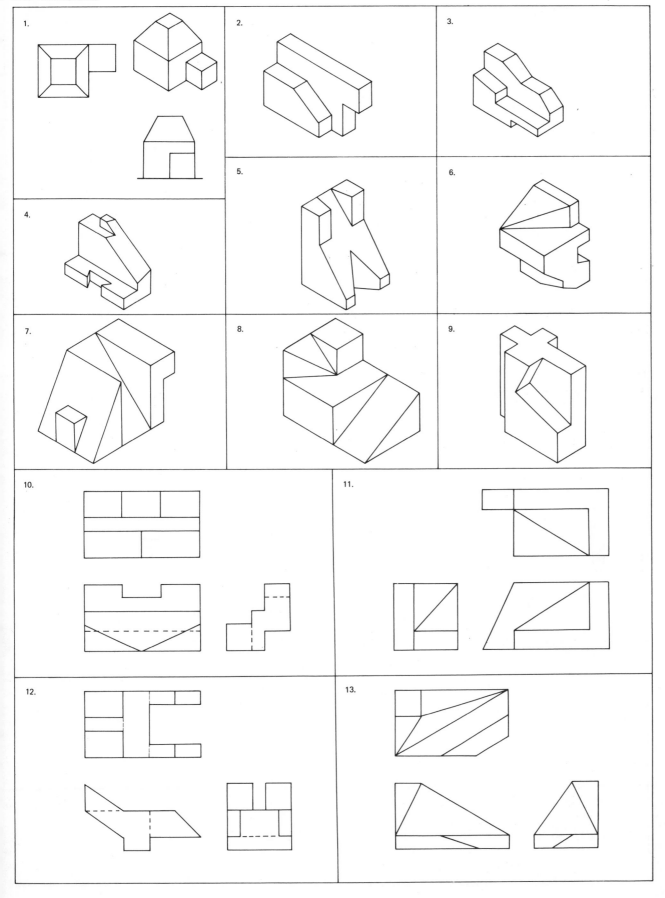

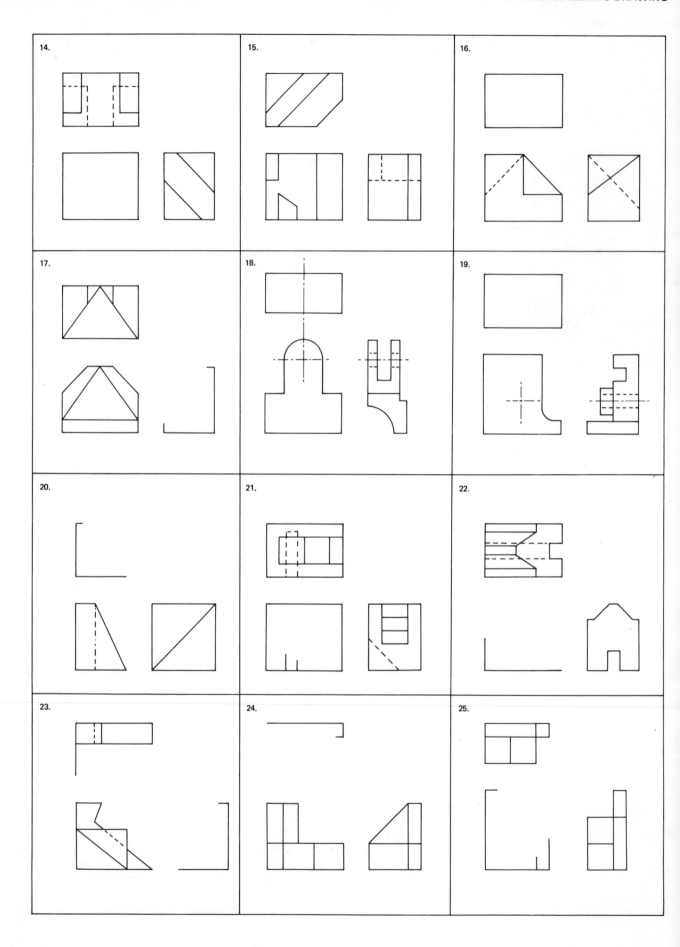

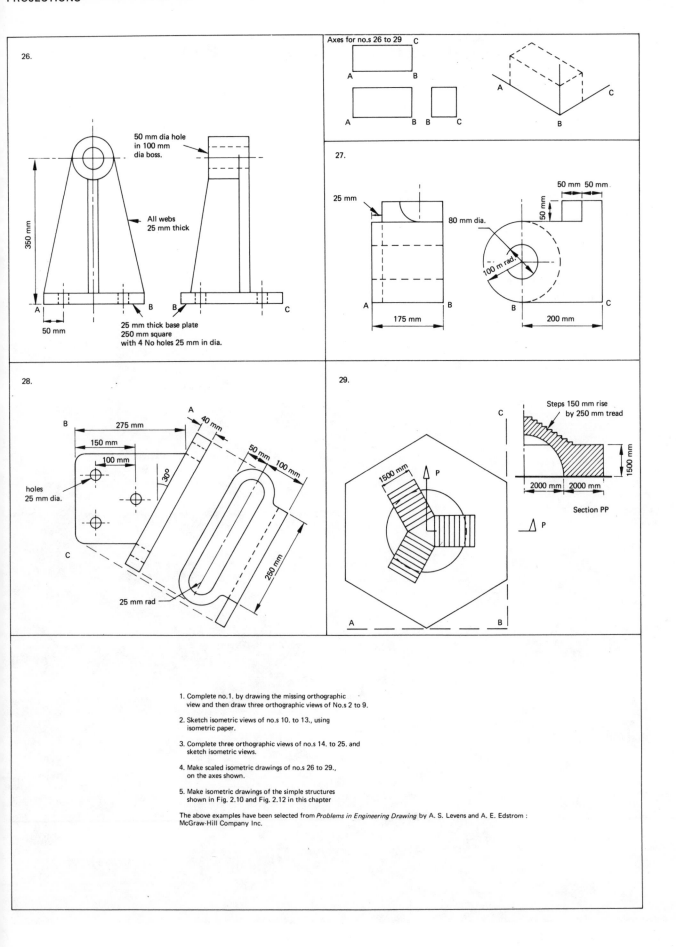

26.

50 mm dia hole
in 100 mm
dia boss.

350 mm

All webs
25 mm thick

50 mm

A   B   B   C

25 mm thick base plate
250 mm square
with 4 No holes 25 mm in dia.

Axes for no.s 26 to 29

27.

25 mm

80 mm dia.

100 m rad.

50 mm 50 mm

50

A   B   B   C

175 mm   200 mm

28.

B   275 mm   A   40 mm

150 mm

100 mm   50 mm   100 mm

300

holes
25 mm dia.

250 mm

C

25 mm rad

29.

C   Steps 150 mm rise
by 250 mm tread

1500 mm   1500 mm

P

2000 mm 2000 mm

Section PP

P

A   B

1. Complete no.1. by drawing the missing orthographic
   view and then draw three orthographic views of No.s 2 to 9.

2. Sketch isometric views of no.s 10. to 13., using
   isometric paper.

3. Complete three orthographic views of no.s 14. to 25. and
   sketch isometric views.

4. Make scaled isometric drawings of no.s 26 to 29.,
   on the axes shown.

5. Make isometric drawings of the simple structures
   shown in Fig. 2.10 and Fig. 2.12 in this chapter

The above examples have been selected from *Problems in Engineering Drawing* by A. S. Levens and A. E. Edstrom :
McGraw-Hill Company Inc.

Chapter 3   **Conventions and presentation**

An earth dam, Celyn, North Wales. The 40 m high dam, together with Bala Lake, regulates the flow in the River Dee to prevent flooding and maintain higher minimum flows so that water can be abstracted all the way down to Chester for domestic and industrial use. (Photograph supplied by the Water Engineer, City of Liverpool Water Department)

3.1 INTRODUCTION

Conventions range from the type of line and printing to be used on a drawing, through formalized shading of sections, to the widely used and understood conventional signs of *Ordnance Survey* maps.

Presentation is concerned with the overall appearance of the finished drawing. It can be influenced by the position of the views within the available space on the paper, the thickness and variety of the lines used, the type of printing, etc. An engineer will leave something of himself in each of his drawings; his style and presentation will be different in detail from that of his colleagues. Some people have a happy flair for layout, while others struggle hard and achieve less elegant results.

A few of the more important conventions and details of presentation are considered in this chapter. Many of the items are further described in the three British Standards, BS 308: *Engineering Drawing Practice*, BS 1192: *Drawing Office Practice for Architects and Builders*, and BS 3429: *Specification for Sizes of Drawing Sheets*. Where possible a direct quotation or illustration from the appropriate British Standard is used.

3.2 SIZES OF DRAWINGS (BS 3429)

The metric or A sizes of drawings are listed in Table 3.1, with reference to some of the approximate Imperial sizes with their resounding names.

The basic metric size, A0, is a rectangle of one square metre with its sides in the ratio of $1: \sqrt{2}$. Derived sizes are either successively halved or doubled, but still retain the same ratio of long side to short side. In this way, smaller sheets can be made from larger ones without waste. This was not always possible with the old Imperial sizes.

The A1 size is slightly larger and, therefore, more convenient to use, than the Imperial size formerly used by most students. The A4 size is used for reports and lecture notes.

Most engineering drawing will be on the A0 size, which replaces the popular double elephant (30 x 40 in.) and the longer antiquarian (30 x 53 in.).

The larger the drawing the smaller the number of drawings required to cover a given scheme. There is a move towards smaller, more precise drawings which are more convenient to handle. Engineers, who have had to manage large drawings on a windy site in the rain or in a car, will appreciate why this movement has started.

Strip plans of roads or rivers, with their associated cross sections, are more convenient on small drawings, A2, A3, or A4, clipped together in album form. Moreover, there is less chance of the drawings being damaged in use.

Although many existing drawings could quite well be presented serially on small sheets, the common procedure is to use larger scale ratios and more precise drawing techniques. Many large drawings waste space which represents time and money.

Amendments can be made more easily by redrawing a whole, small sheet rather than by laboriously erasing large parts of a conventional drawing. Rush jobs can flow from the drawing office in a fairly steady stream of small drawings rather than in gulps as they often do when the more conventional method is used.

3.3 LAYOUT

It is recommended that the standard layout for contract and working drawings should be as shown (in Fig. 3.1).

It is recognised that latitude may be necessary in the arrangement of title and revision panels, notes, etc., to meet differing requirements, but it is considered that a standard layout of sheet will facilitate the reading of drawings and make it possible for essential references to be located easily, especially when drawings are prepared by several offices. A standard arrangement tends to ensure that all necessary information is included. (BS 1192)

A particular firm may have good reasons for keeping its established drawing layout, but

4 A0	1682 x 2378 mm			
2 A0	1189 x 1682 mm	Hamburg	(40 x 60 in.)	1015 x 1530 mm
A0	841 x 1189 mm	Double elephant	(30 x 40 in.)	763 x 1015 mm
A1	594 x 841 mm	Imperial	(20 x 30 in.)	508 x 763 mm
A2	420 x 594 mm			
A3	297 x 420 mm			
A4	210 x 297 mm	Foolscap	(8 x 13 in.)	203 x 330 mm

Table 3.1

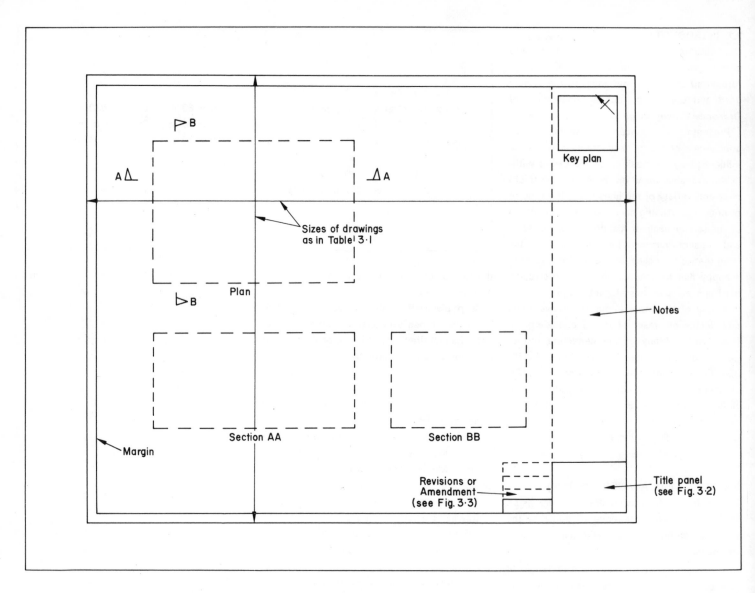

3.1 *General layout of a drawing sheet show-*
ing margins, title panel, amendment
block, and the space generally reserved
for notes. With the suggested location
of plan and sections.

most requirements will be met by the standard layout shown in Fig. 3.1. This layout is suitable for drawings which will be kept horizontally in drawers, since the title panels, etc., can be referred to without having to remove the drawings. You should adopt this layout for all of your exercises, using a simplified title panel at first.

Drawings can also be filed vertically which usually saves floor space and gives better access to each drawing. Some vertical systems have punched suspension strips that are attached to the drawings so that they can be hung by their top edge. For this arrangement, the title panel will have to be at the top of the drawing so that it can be seen while the drawing is still in the cabinet. Some vertical filing systems can store conventionally titled drawings and still allow the title panel to be seen.

The standard layout should be maintained even for 'upright' drawings, that is where the vertical axis of the drawing is parallel to the long sides of the sheet.

Most firms standardize their work to two or three sizes of drawing. It is better to waste some space on a standard drawing rather than to be faced with filing an odd sized drawing which will either slip to the back of a drawer or be a nuisance in a vertical cabinet. The draughtsman's time is saved if printed sheets are used. For routine work, the printing consists of a border and title panel with the name of the firm or authority. A more complete title panel should be provided for large schemes for which many drawings will be made.

(a) Title panel

A newcomer to any drawing will need to assimilate a minimum of information as quickly as possible before he can interpret

(a) Client	**DALESFORD R.D.C.**
(b) Scheme	LONGBOTHAM SEWAGE SCHEME
(c) Part of scheme	**PLAN and SECTIONS of No 3 BOOSTER PUMPING STATION**
(d) Engineer or consultant	**ROBINSON and PARTNERS** Chartered Engineers **743 VICTORIA STREET LONDON S.W.I.**

| *Scales*: 1:100 and 1:50 |
| *Date*: JUNE 1970 |
| *Dwg, No.* 674 C |

| *Drawn*: T.B.S. | *Checked*: S.B. | *Traced*: C.F.B. |
| | *Approved*: J. Robinson |

the drawing. Such information is best presented in the title panel which should always be located in the same place on each drawing.

A cursory glance at a good title panel should tell the observer:

The name of the client or authority for whom the work has been prepared.

The name of the scheme or job.

The part of the job shown on the drawing.

The scales used.

When the drawing was completed.

The name of the engineer or consultant.

The drawing number.

All this information is contained in the title panel shown in Fig. 3.2. Alternative layouts are shown in BS 1192.

The initials of the design engineer are entered in the space marked DRAWN. Civil engineering design, like any design, is open-ended, i.e., there is no single solution. As a result, the chances for errors to creep in are many. It is, therefore, unfair to lay all the responsibility for the design on one man. In an attempt to uncover errors, an independent check of the work must be carried out by another engineer, preferably one who has had little or nothing to do with the detail of the design concerned. His unprejudiced eye will often discover errors or misconceptions that the designer has lived with and overlooked as the design developed. The engineer who carries out the check enters his initials in the CHECKED space.

The ultimate responsibility for the correctness and feasibility of the ideas expressed in the drawing lies with the chief engineer of an authority or a partner in a consultancy. His name will appear in the APPROVED space.

(b) Revisions or amendments

Few designs remain unmodified between design and construction. An engineer should never be afraid to amend a design at any stage, if it is necessary to do so. Alterations in the construction phase, however, can be

3.2 A typical title panel.

7·12·70	C	Intake lowered: pump supports lowered: rising main extended	Ro
18·11·70	B	Crane details added	7
16·9·70	A	Floor finishes added	
DATE	REF.	DETAILS	Draw
		Amendments	

3.3   Amendment or revision block showing how it grows upwards.

3.4   Suitable north points.

Table 3.2

very expensive.

Alterations, however small, must be recorded in the revision or amendment panel, which grows upwards, see Fig. 3.3. It is a good idea to incorporate the amendment reference in the drawing number so that superceded drawings can be spotted more easily, e.g., drawing 674C obviously supercedes drawing 674A, etc. The amendments or fresh drawings made on completion of the works to record them 'as built' should have a suitable suffix, R (Record), for example.

(c) North points

The orientation of location and site plans is shown by a north point, see Fig. 3.4.

Use a simple but attractive north point.

3.4 SCALES

Drawings for civil engineering range from continental maps with a scale of several millions to one to full-size detail or even larger.

A range of suitable scales and their uses is listed in Table 3.2. The imperial scales they replace are shown alongside.

Scale ratio	Use	Previous equivalent
1: 50 000	Location maps	1 in. to 1 mile 1: 63,360
1: 10 000		6 in. to 1 mile 1: 10,560
1: 2500	Site maps	1: 2500, about 25 in. to the mile
1. 1250		1: 1250, about 50 in. to the mile
1: 500	Site plans	
1: 200	General arrangement	about $\frac{1}{16}$ in. to 1 ft
1: 100		about $\frac{1}{8}$ in. to 1 ft
1. 50		about $\frac{1}{4}$ in. to 1 ft
1: 25	Plans, elevations, and sections	about $\frac{1}{2}$ in. to 1 ft
1: 10		about 1 in. to 1 ft
1: 5	Details	about quarter full size
1: 2		half full size
1: 1		full size

Type of line		Example	Application
Continuous (thick)	A	————————	Visible outlines.
Continuous (thin)	B	————————	Dimension lines. Projection or extension lines. Hatching or sectioning. Leader lines for notes. Outlines of revolved sections.
Short dashes (thin)	C	- - - - - - - -	Hidden details. Portions to be removed.
Long chain (thin)	D	—— · —— · ——	Centre lines. Path lines for indicating movement. Pitch circles.
Long chain (thick)	E	—— · —— · ——	Cutting or viewing planes.
Short chain (thin)	F	— · — · — · —	Developed or false views. Adjacent parts. Feature located in front of a cutting plane. Alternative position of movable part.
Continuous wavy (thick)	G	～～～～～	Irregular boundary lines. Short break lines.
Ruled line and short zig-zags	H	—／—／—／—	Long break lines.

3.5   The types of line suggested in BS 308.

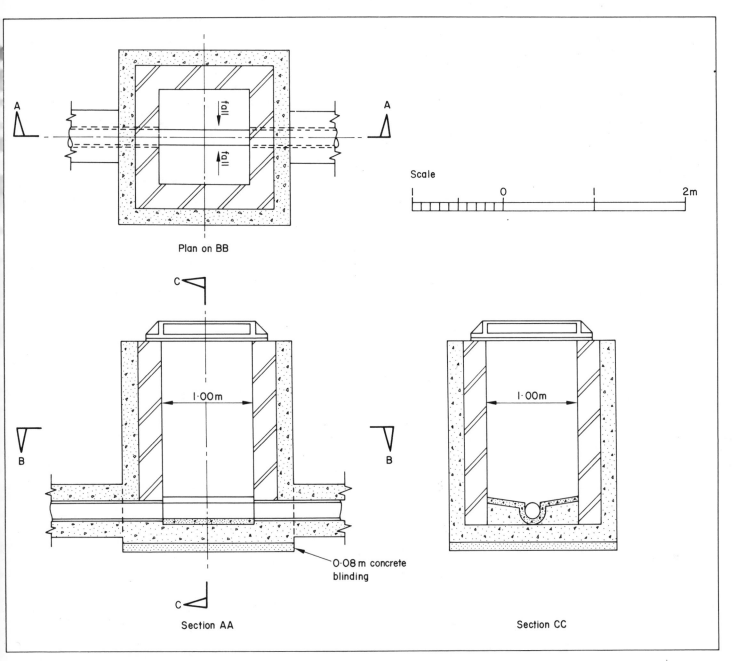

Plan on BB

Scale

0   1   2m

Section AA

0·08 m concrete
blinding

Section CC

Scales, other than those in Table 3.2, should not be used unless there is a very good reason for doing so. It is better to have a drawing that does not quite fill a standard sheet than to have either a peculiar scale ratio or an odd sized drawing.

It is not usual for dimensions to be scaled from a civil engineering drawing as errors will occur due to paper instability. Probably the only exceptions to this rule are drawings of rough earthworks, when it is useful to include a drawn scale, which will have altered with the paper, see Fig. 3.6 and Fig. 3.10. A drawn scale, and no other, must be used for drawings that will be recorded on microfilm or which may be reduced or enlarged by other photographic means.

3.5 TYPES OF LINE
Lines of similar thickness throughout make for a dull drawing, while careful grading of line thickness, with perhaps the skilful use of an extra heavy line here or there, adds interest. This is a good reason for you to start using pens at once for it is difficult to vary the thickness of pencil lines.

The basic types of line, as suggested in BS 308, are shown in Fig. 3.5.

Lines should be sharp and dense to obtain good reproduction.

3.6 *Arrangement drawing of a small sewerage manhole which shows several types of line, some conventional shading, a drawn scale, and how to reference sections.*

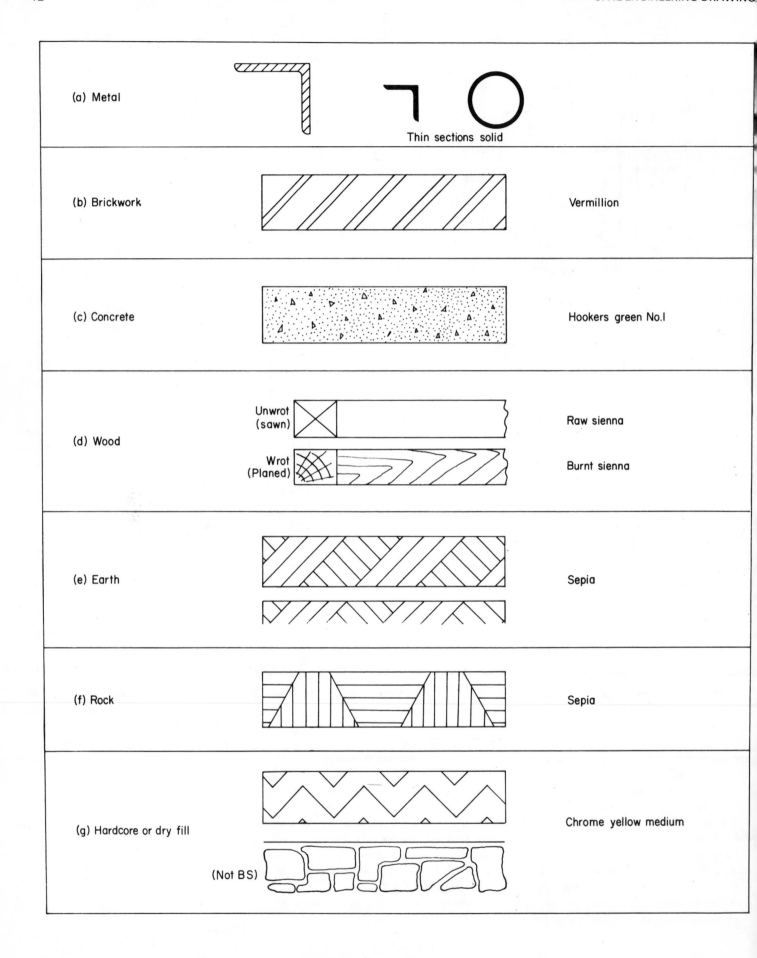

(a) Metal   Thin sections solid

(b) Brickwork   Vermillion

(c) Concrete   Hookers green No.I

(d) Wood
Unwrot (sawn)   Raw sienna
Wrot (Planed)   Burnt sienna

(e) Earth   Sepia

(f) Rock   Sepia

(g) Hardcore or dry fill   Chrome yellow medium
(Not BS)

Lines specified as thick should be from two to three times the thickness of lines specified as thin.

Centre lines should project for a short distance beyond the outline to which they refer, but where necessary to permit dimensioning, they may be extended as dimension lines (Type B). Centre lines should not intersect in the spaces between dashes.

Lines depicting hidden details should always begin and end with a dash in contact with the visible or hidden detail line at which they start or end, except when such a dash would form a continuation or a visible detail line. Dashes should join at corners and arcs should start with dashes at the tangent points. (BS 308)

Various types of line are used in Fig. 3.6 together with some of the sectioning conventions of Fig. 3.7.

3.6 SECTIONING

The methods of showing some of the common structural materials when they are exposed in sections are shown in Fig. 3.7.

Although drawings are not often coloured today, a few, carefully chosen, light washes of the conventional colours on a section will enliven the drawing and help the lay observer in his interpretation. The same colour is used for elevational and sectional views of a material, but the section is made darker by the application of a second or third wash after the previous washes have dried.

Stick-on shading and other details

A variety of special shading effects can be obtained by using self-adhesive transparent material. It is suitable for sticking on tracings, to indicate shading for sections, north points, drawn scales, etc. The use of stick-on shading speeds the draughting process, while its uniformity creates a more professional effect. Although many of the shading patterns are prepared essentially for commercial artists, the engineer finds some of them useful either on conventional drawings or on cartoons.

It is often more convenient to apply stick-on material to the back of a tracing so that it does not interfere with the linework of the drawing.

Transparent, sensitized, self-adhesive material is available for the preparation of special stick-on shapes. It is a dyeline material, see section 9.4, which can be printed from any negative and is developed conventionally. It has a siliconized backing sheet which is removed to reveal the self-adhesive surface. It is best given a reversed image so that it can be stuck on the back of the tracing. It can be used for standard title panels or, in order to save time, for inserts of maps taken from transparent *Ordnance Survey* sheets.

A selection of stick-on sectioning and shading and other details is shown in Fig. 3.8.

3.7 DIMENSIONS AND LEVELS

A particular form of dimension met in civil engineering is the level, i.e., the height of a point above a chosen datum which is usually the *Ordnance Survey* datum, but can be any convenient reference. Levels replace vertical dimensions on drawings which leads to simplification.

Figure 3.9 (a) shows the system of dimensions suggested in BS 308 which is applicable to civil engineering practice.

The nature of civil engineering work allows most dimensions to be given in multiples of 10 mm, i.e., 10.47 1045.00 or 0.76 in metres. If the scale of the drawing or the resolution require it, dimensions can be given in millimetres i.e., 765, 1574 or 5. The presence of the decimal point will indicate whether the units are metres or millimetres. Dimensions must be consistent on a drawing, wholly in metres or millimetres, not mixed. An occasional dimension shown with a resolution closer than 10 mm, say 10.473, will draw attention to a particular need for greater accuracy. Levels can comfortably be read to 5 mm, i.e., 104.675; accuracy closer than this is often wishful thinking.

Figure 3.9 (b) shows some frequently used alternatives to the arrowheads used in (a). Arrowheads should be slim, say 5° each side of the shaft, and can be filled in or left hollow.

Dimensions can sit on the dimension line or be let into it. The two systems should not be mixed in one drawing. Dimensions on a vertical dimension line should have their base to the right, the natural way a right-handed draughtsman would write them. The left-hander will either have to remove the drawing from the board and twist it into a comfortable position or walk round the board, if he can, to print the figures.

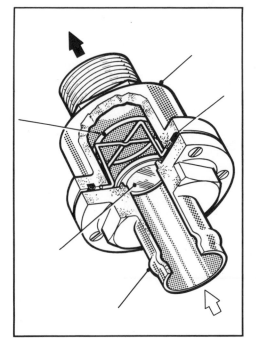

3.7 *Conventional shading and colours for some constructional materials (BS 1192).*

3.8 *Examples of some stick-on symbols and shading (some are special orders). (Photograph supplied by Letraset Limited.)*

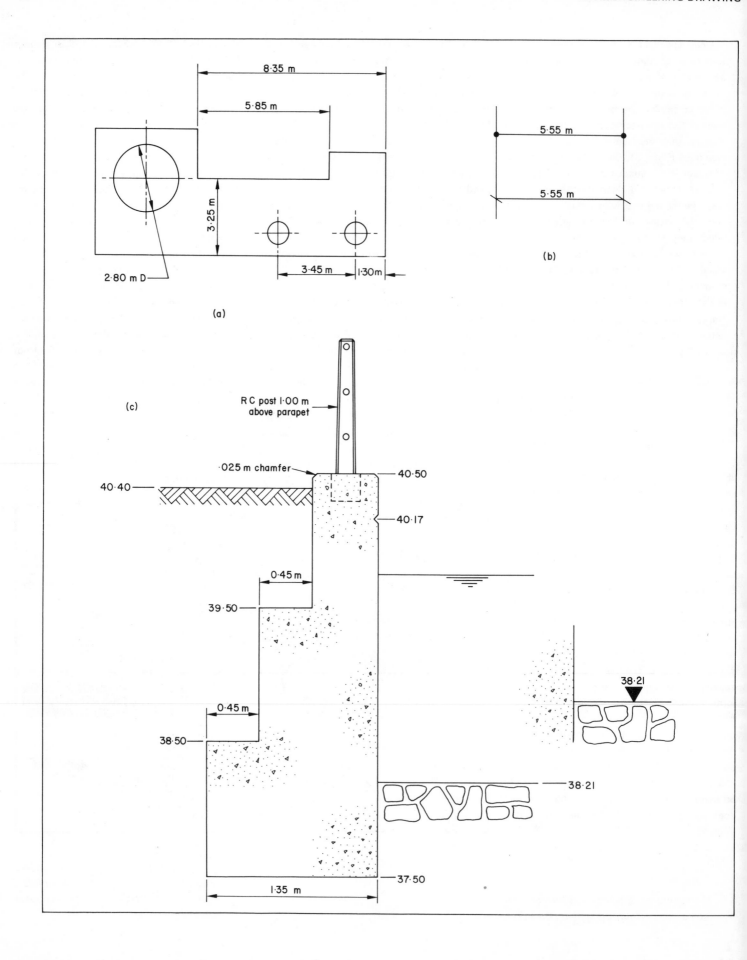

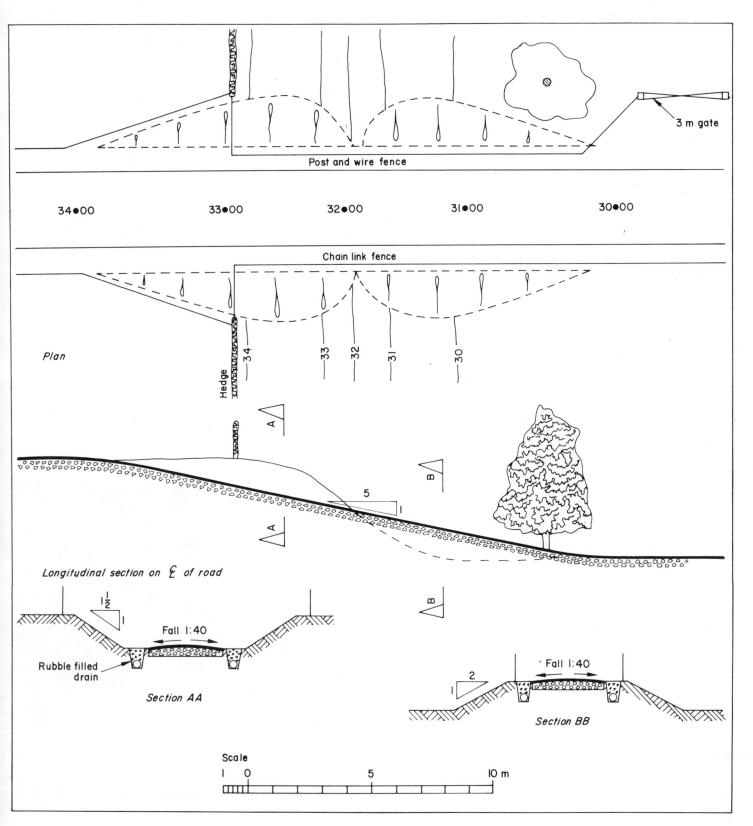

3.9 Methods of showing dimensions.

    (a) From BS 308.

    (b) Alternatives sometimes used for arrowheads.

    (c) Two methods of showing levels.

3.10 Simple earthworks with several methods of showing slopes, contours, hatching (tadpoles), spot levels, etc.

Figure 3.9 (c) shows a suitable system of dimensioning for civil engineering work. A shaded triangle to indicate levels is acceptable when a leader line cannot conveniently be taken to a clear part of the drawing, e.g., invert and soffit levels. (The invert of a pipe or tunnel is the bottom where the water flows, the soffit is the underside of the roof.)

The level of a flat area in a plan, a concrete deck, for example, is shown in a rectangular box, e.g., ⟨37.35⟩ Some authorities also suggest boxing the values of levels in sections and elevations as well. A spot level on a road or in a field is shown as a dot, with the value written beside it. It is often convenient to use the decimal point of the level as the dot, see Fig. 3.10.

A contour line, which is a line in the plan joining points of the same altitude and usually restricted to natural slopes, is also shown in Fig. 3.10.

3.8 SLOPES, FALLS, BATTERS, AND GRADIENTS

These are all words the engineer uses (sometimes indiscriminately) for a slope.

Batters are usually the steep faces of embankments or cuttings, see Fig. 3.10, while gradients are the less severe slopes of roads or railways.

Fairly steep slopes are shown in the plan by tadpoles which swim uphill, while flatter slopes are shown by contour lines. Tadpoles are usually restricted to artificial slopes the limits of which can be shown by dotted lines.

3.9 LETTERING

Many good drawings are ruined by poor printing.

You should aim at a simple style that can be executed quickly freehand, see Fig. 3.11 (b). BS 308 suggests the types of lettering shown in Fig. 3.11 (a).

Do not make the letters too big.

Use a slightly softer pencil which is not too sharp.

Few of us can make a really neat job of letters much more than 10 mm high, so it is sometimes worthwhile using letter stencils for titles, but this is very slow. For prestige work, the very professional effect obtained by stick-down letters such as *Letraset, Presletta*, etc., is worth the extra time and effort involved, see Fig. 3.11 (c).

(a)

(b)

(c)

3.11 Some examples of printing.
   (a) BS 308.
   (b) Freehand.
   (c) Self-adhesive transfers. (Photograph supplied by Letraset Limited.)

Part Two   **Geometry**

Chapter 4   # Points, lines, surfaces, and shadows

Water transport, a shipbuilding dock for Messrs Harland & Wolff in Belfast. The dock is 550 m long and will accommodate ships up to one million tons deadweight. Ships will be launched by flooding the dock. Shipbuilding started at one end while the dock was still being built at the other. (Photograph supplied by George Wimpey & Co., Ltd.)

4.1 INTRODUCTION

The engineer can conveniently divide the surfaces of three-dimensional shapes into two groups, those which can be developed precisely and those which cannot.

A surface is said to be developed when it is laid out flat. For example, the faces of the cube on to which the orthographic views in Chapter 2 are projected are developed in Fig. 2.4, although only three of the six faces are shown. The resulting flat shape is called the development of the surface.

Plane or singly curved surfaces (see definitions below) can be developed exactly, whereas doubly curved or warped surfaces can be developed only approximately.

Figures 4.1 and 4.3 show how several different surfaces are formed by a moving line called a generator. Any position of the generator is an element of the surfaces.

A plane surface is generated, see Fig. 4.1 (a), by a straight line AB moving in a straight line while remaining parallel to its original position. A' B' is an element of the surface.

A singly curved surface is generated, see Fig. 4.1 (b), by a straight line moving so that adjacent positions lie in the same plane, e.g., in (bi) A' B' A'' B'' lie in the same plane as AB' AB'' in (bii).

Adjacent positions of the generator of warped surfaces, see Fig. 4.1 (c), do not lie in the same plane. They cannot be developed. Three common examples of warped surfaces are shown.

(ci) A smooth transition between trapezoidal and rectangular channels.

(cii) A parabolic hyperboloid in which the stiffness of a doubly curved structure is achieved with straight elements A' B'.

(ciii) A circular hyperboloid whose generation is best understood by imagining straight elements attached to circular plates top and bottom which are then rotated relative to each other, 90° in this case.

Although (cii) is often known as an hyperbolic paraboloid, it is better to think of all warped surfaces which have the hyperbola as the main contour as hyperboloids. Sections other than the main one will reveal other conic sections, see Chapter 5, which classify the shape within the hyperboloid family. The cooling tower or circular hyperboloid in (ciii) has circular horizontal sections.

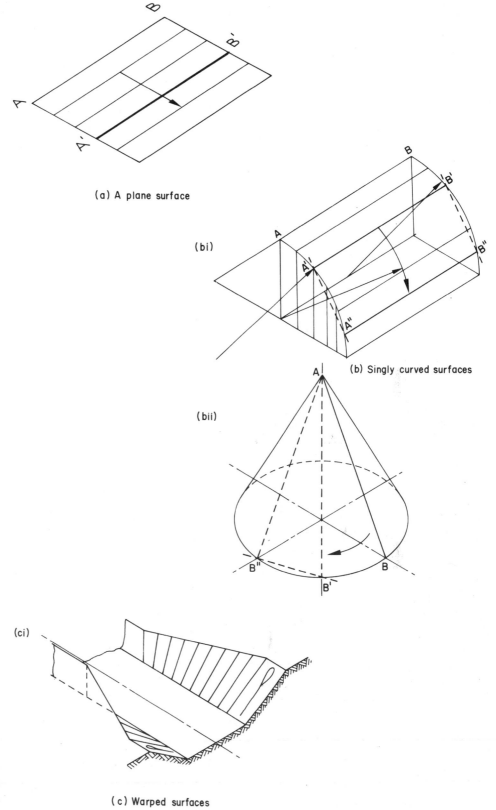

(a) A plane surface

(bi)

(b) Singly curved surfaces

(bii)

(ci)

(c) Warped surfaces

4.1 Generation of surfaces.
   (a) Plane surface by a straight line.
   (b) Singly curved surfaces by a straight line.
   (c) Warped surfaces by straight lines.

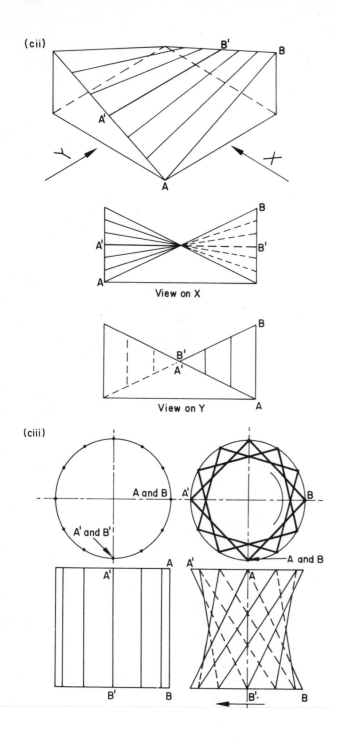

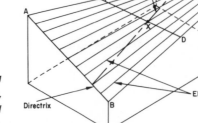

4.2 Construction of an hyperboloid showing elements touching all three directrices, AB, CD, and EF, with an additional element passing through point X on CD.

The Americans refer to hyperboloids as double-ruled surfaces because two elements can pass through certain points on their surface.

An hyperboloid is defined, see Fig. 4.2, by three nonparallel directrices, AB, CD, and EF, that do not intersect. The surface is generated by elements that touch all three directrices, the lines that direct or determine the shape.

The circular hyperboloid has circular directrices, one at the top, one at the bottom, and the waist or gore circle. Two different elements can cross on the middle directrix.

If smooth warped surfaces are to be built in concrete, the shuttering can be made of planking laid parallel to the elements of the surface with a minimum of twist in each board. While the necessary cables, if prestressed concrete is being used, can be straight with consequent easing of the tensioning process.

It is sometimes convenient to imagine that a singly curved surface is generated by a curved line moving in a straight path, see Fig. 4.3 (a).

Doubly curved surfaces are generated, see Fig. 4.3 (b), by curved lines moving on curved paths, at a radius R (bi) or about its own axis (bii).

4.2 LINES IN AN ENGINEERING DRAWING

Lines in an engineering drawing represent:
- (a) The edge view of a plane.
- (b) The intersection of two planes.
- (c) The horizon of a curved surface.

Compare the orthographic view in Fig. 4.4 with the isometric one while considering (a), (b), and (c) above.

- (a) The line CA represents the edge of the plane CAJG.
- (b) The line BA represents the intersection of planes BAC and BAD.
- (c) The line EH represents the horizon of the curved upper block when seen in the direction of the arrow X, and also the edge of the plane EFGH. Note that there is no line PQ as there is no sharp edge there.

The single orthographic view in Fig. 4.4 does not fully describe the shape of the block since there is no indication of the curved surface. A view in the direction of the arrow Y would show the curvature, but there could be doubt about the fillet.

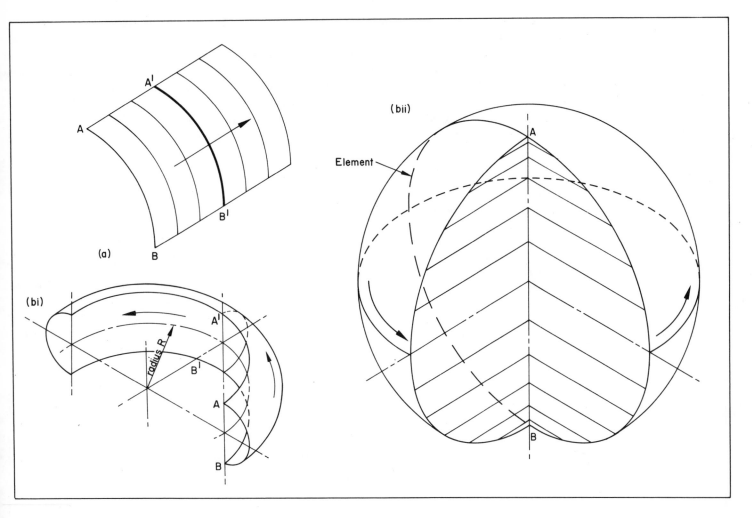

(a)

(bi)

(bii)

Element

radius R

4.3 Generation of surfaces.
(a) Singly curved surface by a curved line.
(b) Doubly curved surfaces by a curved line.

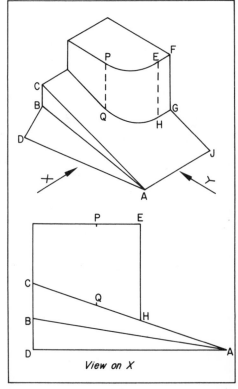

View on X

4.4 Use of lines in engineering drawing to show.
(a) The edge of a plane—line CA.
(b) The intersection of two planes—line BA.
(c) The horizon of a curved surface—line EH.

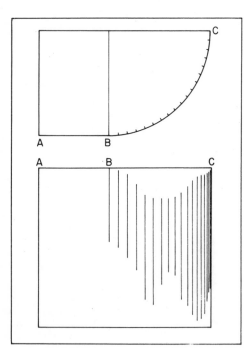

Sometimes, the impression of a curved surface can be given nonconventionally by light shading. In Fig. 4.5, equally spaced points located round the curve in the plan, when dropped into the elevation, become closer from B to C thereby inferring a curved surface bending into (or out of) the plane of the paper.

4.5   Showing curved surfaces nonconventionally.

4.3 DESCRIPTIVE GEOMETRY

Descriptive geometry is the graphical solution of problems concerned with points, lines, and surfaces in three dimensions and is intimately concerned with engineering drawing and projection.

(a) Lines

Figure 4.6 (a) is a three-dimensional view

4.6   Line AB on the sloping face of a wedge.
*   (a) Three-dimensionally.*
*   (b) Orthographically with auxiliary section to find the true length and inclination of AB.*

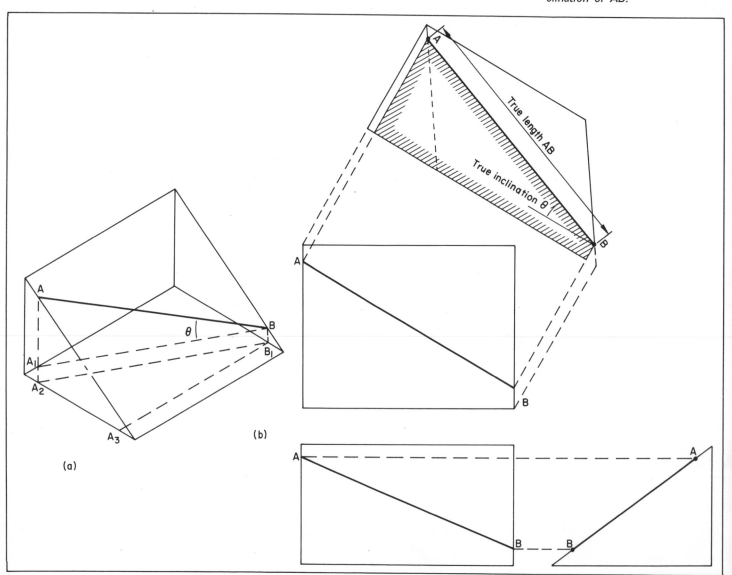

of a line AB on the sloping face of a wedge. AB is inclined to all three planes of projection.

The first step is usually to find the true length of the given line and its inclination to the horizontal.

This can be done analytically from the co-ordinates of points A and B using Pythagoras' theorem to find in turn the length of the hypotenuse A_2B_1 of triangle $B_1A_3A_2$ and then AB from triangle AA_1B. As all the sides of triangle AA_1B are known, the angle θ can be found.

Figure 4.6 (b) shows three orthographic views of the wedge with the line AB, which is foreshortened in each view. An auxiliary view is necessary, projected on to a plane parallel to the line AB in the plan, to show the true length of AB and its inclination θ to the horizontal.

The centre line of a road, with a constant gradient, climbing up the side of a valley, is an application of a line on an inclined plane. With contour lines 5 m apart, as shown in Fig. 4.7, and a road gradient of 1 vertical in 10 horizontal, successive points, 5 m apart vertically, will have a horizontal separation of 50 m. Starting at A, with dividers set to 50 m, points can be located on successive contour lines by walking the dividers up the slope. As the gradient is so flat (geometrically speaking, although it is sufficient to slow down heavy vehicles), the true length of the road is almost the same as its plan length; each 50 m horizontal intercept represents a piece of road $50 \times \sqrt{\dfrac{101}{100}}$ m long.

(b) Planes

Figure 4.8 shows how an infinite plane can be defined by three points A, B, and C.

The determination of the true slope of a plane is an essential exercise in geology where the plane is represented by a plane stratum of rock and the true slope is known

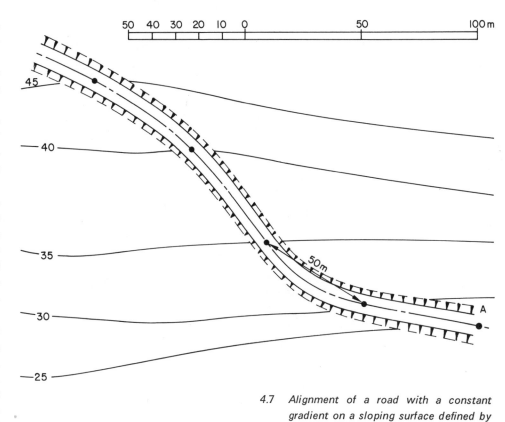

4.7   Alignment of a road with a constant gradient on a sloping surface defined by contour lines.

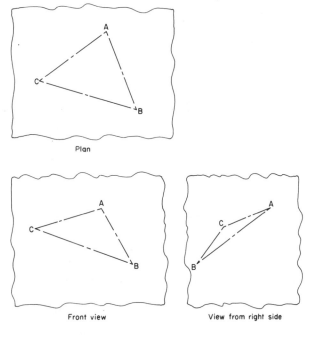

4.8   Orthographic views of an inclined infinite plane defined by the three points, A, B, and C.

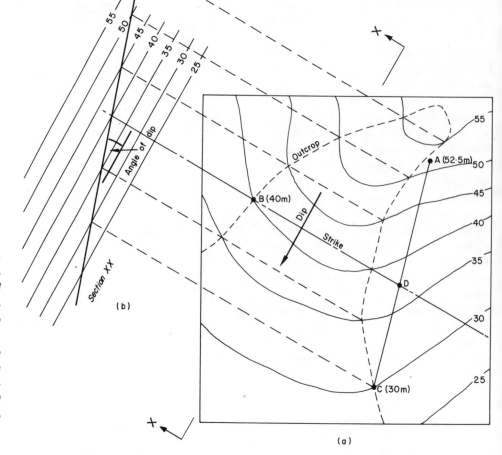

(a)

4.9 (a) *Contour map of three outcrops, A, B, and C, of the same rock stratum, showing location of D at same level as B to define the strike. The inferred shape of the outcrop is also shown.*

(b) *Section XX taken at right angles to the strike to show the dip or true slope of the stratum, with construction lines for determining points of the outcrop on contour lines in the map.*

as the dip of the stratum.

Figure 4.9 (a) is a contoured map showing three outcrops of rock A, B, and C, which are known to be on top of the same plane stratum of rock. First find the strike of the stratum, which is the bearing of a horizontal line on its surface, by finding the point D in AC which lies at the same level as B, by dividing AC proportionately. The dip lies at right angles to the strike. Its value can be found graphically by drawing the section XX, see Fig. 4.9 (b), as an auxiliary view on a plane parallel to the dip.

Once the dip is known, the shape of the outcrop can be found on the map. Points on the outcrop are located graphically where lines from the section, projected parallel to the strike, intersect contour lines with the same level.

4.4 DEVELOPMENTS

A surface is said to be developed when it has been laid out flat; the resulting flat shape is called the development of the surface.

The civil engineer may become involved with ducting for ventilation or with fabricated

pipes and tunnels for hydraulic schemes. Although he may not be concerned with the fabrication of pipes, a knowledge of what is or what is not readily developable will result in a better, cheaper job for his client.

Most civil engineers, however, will be concerned with shuttering for concrete at some time in their careers, and for this, a knowledge of development techniques is essential, either in the design and erection of the shuttering or in choosing seemingly complicated shapes that can be shuttered simply.

Developments are closely connected with intersections, see section 4.5. Both are exercises in finding the true length of lines in space inclined to the orthographic axes.

(a) Development of a right circular cone

Figure 4.10 shows the plan, elevations, and development of a right circular cone. A singly curved and, therefore, developable surface which is created by the movement of the generator OG so that O remains fixed while G describes a circle. Its instantaneous position OG is an element of the cone.

The development of the cone is a segment

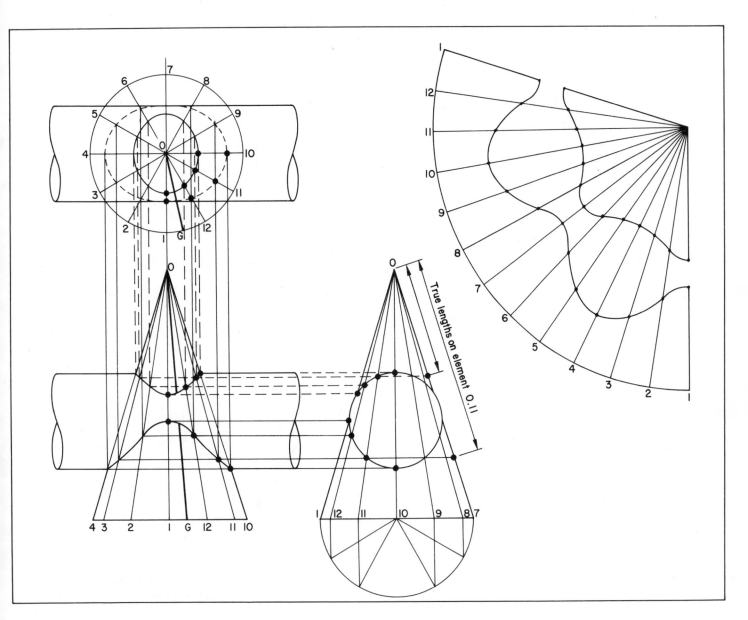

of a circle with a radius equal to the true length of the generator and a peripheral length equal to the length of the path traced by G in the plan. In Fig. 4.10 the length 1.1 in the development is twelve steps of the chord be_ tween adjacent numbered points in the plan. It is convenient to use a 'sixty thirty' set square to divide the plan. In full size work, the periphery of the development would be calculated accurately and then be set out carefully on the curve.

The elemental method of developing a cone can very quickly solve problems of penetration of the cone by other geometric shapes, such as by the cylinder in Fig. 4.10.

The heavily marked intersections of the cyl- inder with the numbered elements of the cone in the side elevation are projected into the

front elevation and, then, up into the plan. Both front elevation and plan can then be completed.

The true length of the generator OG appears only at O.4 and O.10 in the front elevation and at O.1 and O.7 in the side elevation. Thus true distances of the intersection points from the apex of the cone, required for construction of the development, are found by projecting the intersection points horizontally to the line O.7 in the side elevation, or using the intercepts on lines O.1 in the side elevation or on O.10 in the front view.

(b) Development of an oblique circular cone

An oblique circular cone is shown in Fig. 4.11. It has a circular base to which its axis is inclined.

The true length of all the elements must be

4.10 *Development, by use of elements, of a right circular cone penetrated by a cylinder.*

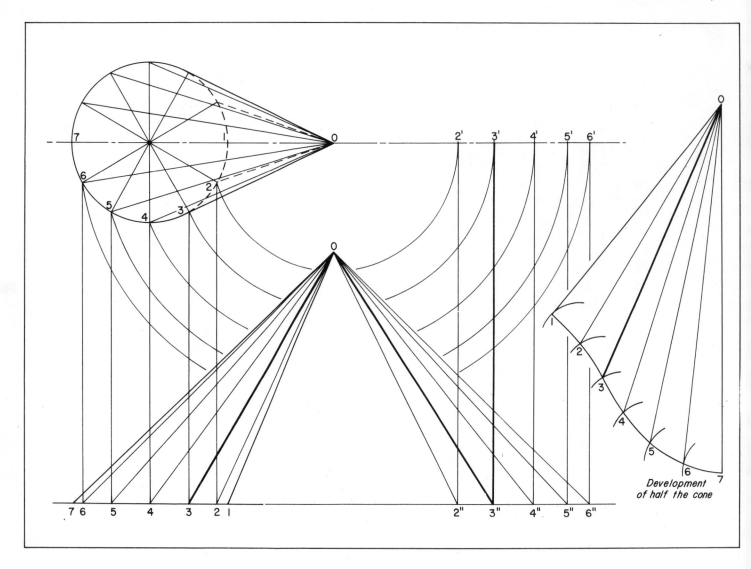

4.11 *Development, by use of elements, of an oblique cone.*

found. O.1 and O.7 alone show their true lengths in the elevation. The horizontal projection of the elements appears in the plan. The elements have a common vertical projection, the height of the cone. The true length of an element will be the hypotenuse of a right-angled triangle incorporating its two projections.

Consider element O.3:

(a) Centre O, radius O.3 in the plan, swing an arc to cut the centre line produced at 3′.

(b) Drop a vertical from 3′ to meet the base of the elevation produced at 3″.

(c) The true length of O.3 appears as O.3″ in the elevation.

To complete the development:

(d) Set up O.7, taken directly from the elevation.

(e) Centre 7, radius equal to the chord 6.7, swing an arc.

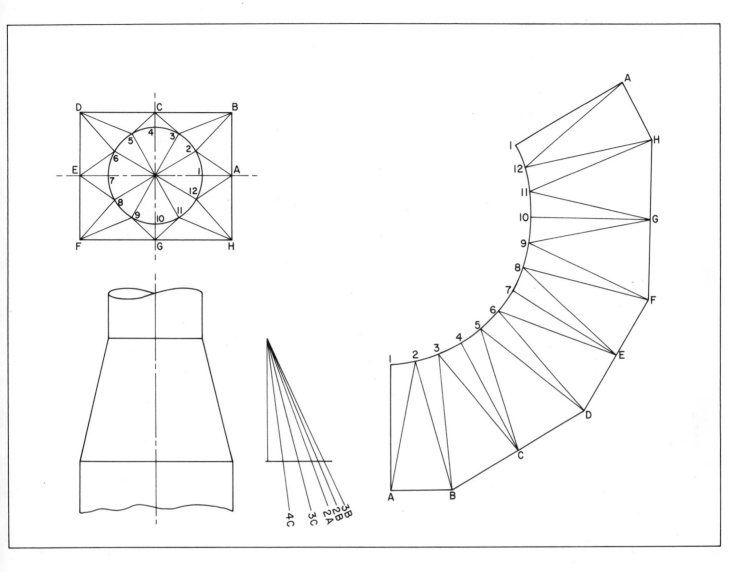

4.12 *Triangulation to develop a compound surface.*

(f) Centre O, radius O.6″, swing an arc to cut the previous arc at 6.

(c) Development of compound shapes

All shapes, whether regular or not, can be developed by the true length element method of triangulation used for the oblique cone in section 4.4 (b).

This method is applied to the transition between a circular duct and a rectangular one in Fig. 4.12.

The chosen section lines or elements which triangulate the transition must pass through salient points, e.g., the corners of the rectangular duct. In the plan, the periphery of the circular duct is divided by radial lines 30° apart. The true lengths of the elements in the quadrant ABC are found in the separate construction; they have a common height and their horizontal projections are transferred from the plan.

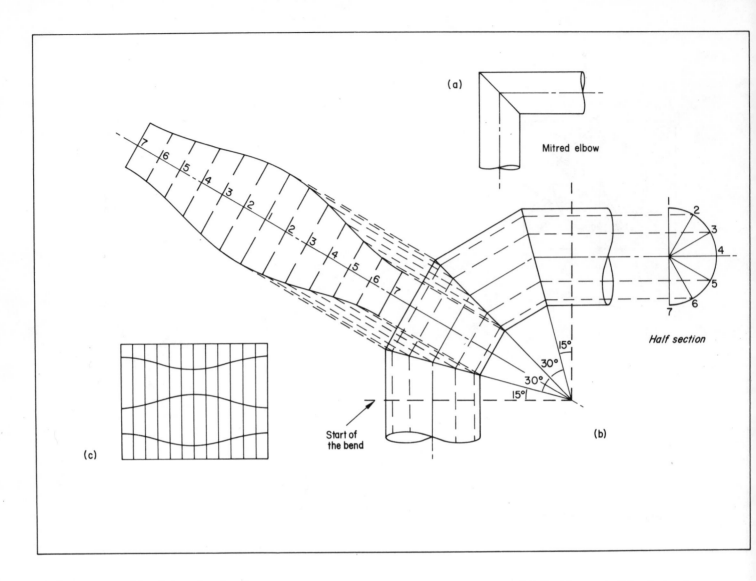

4.13 *Development of bends in a pipe showing the 'lobster backed' elements and how they can be cut with little waste.*

(a) *A mitred elbow.*

(b) *Development of one element, a 'lobster backed' plate.*

(c) *The four elements developed economically on one sheet.*

(d) Development of bends in pipes or tunnels

The simple mitred elbow shown in Fig. 4.13 (a) is very bad hydraulically. A bend made up of two or more segments, however, as shown in (b), is almost as good as a smooth one made of doubly curved surfaces, but it is much easier to make and, therefore, cheaper.

The problem is to find true lengths that can be set out on a plane to construct the development. In this case, the symmetrical development can be defined by two dimensions at right angles without the need for triangulation. The two dimensions are round the periphery of the pipe and along its axis. The former appear in the half section as the chords 12, 23, 34, etc., and the latter appear directly on the plan as the dotted lines constructed from the points 1, 2, 3, etc., on the section.

While the shape of the elements of the bend is best understood as they are developed

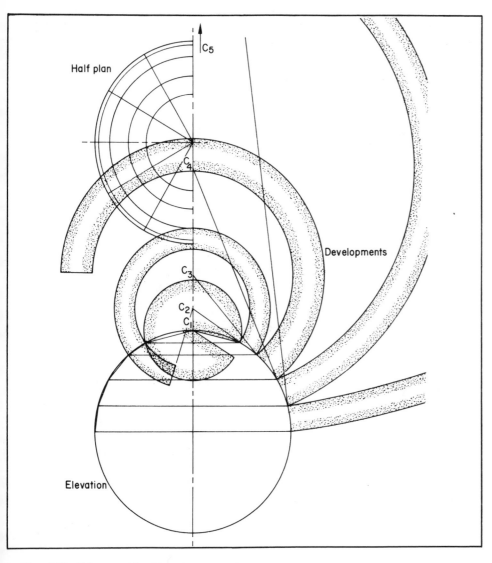

4.14 Approximate development of a doubly curved surface—frustra of cones.

in Fig. 4.13 (b), considerable economy of material and cutting will result from the re-arrangement of the pieces as shown in (c). Moreover greater strength will result after assembly since the longitudinal welds are staggered.

Note that as the ellipses at the end of each completed element must be identical, the first join is made 15° after the start of the bend.

(e) Approximate developments of doubly curved surfaces

The following treatment of the sphere should enable you to devise approximations of other doubly curved shapes.

In Fig. 4.14 the elevation of the sphere is divided into a number of horizontal slices thus creating a conical cap and a number of elemental frustra. Note that each frustum is from a right circular cone of different apex angle and will give rise to circular elements

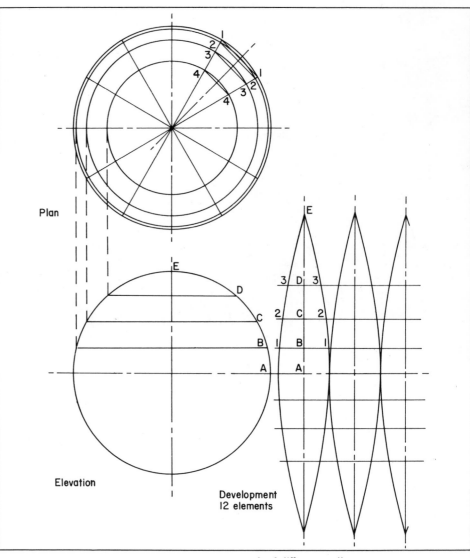

4.15 *Approximate development of a doubly curved surface—segments.*

each of different radius.

Each element can be drawn precisely, but much material is wasted in cutting each one out.

In Fig. 4.15 the elevation is divided into a number of slices and the plan into a number of equal segments. The peripheral lengths of the slices AB, BC, etc., in the elevation represent the lengths of the elements, while the plan lengths 11, 22, 33, represent the widths.

The elements are similar and so can be produced economically in a series.

4.5 INTERSECTIONS (interpenetrations)

An intersection is the junction between geometric shapes, e.g., a hole punched in a card is the intersection of a plane and a cylinder.

While precise developments are limited to plane and singly curved surfaces, orthographic views of intersections of doubly curved sur-

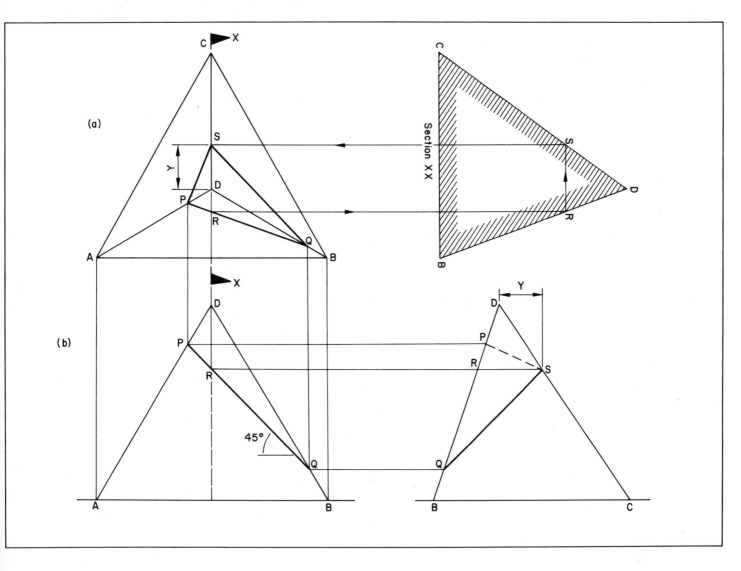

4.16 Intersection of planes.
(a) To form a pyramid.
(b) To intersect the pyramid.

faces can be drawn precisely.

An extended study of a particular inter-section that of a plane with a right circular cone, is made in Chapter 5.

The intersections studied in this chapter range from planes to singly and doubly curved surfaces.

The problem is one of locating a point in space, usually by finding sections of two or more orthographic views that contain the same point.

(a) Intersection of planes

A pyramid is formed by the intersection of planes. The orthographic views of a tetra-hedron, Fig. 4.16, show the straight lines that result from planes intersecting symmetrically.

You should attempt to draw the three orthographic views of this deceptively simple figure before moving on to more advanced intersections.

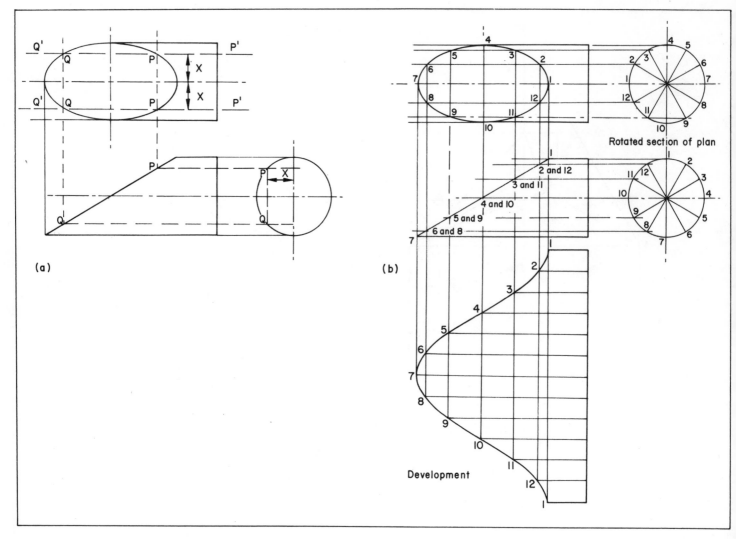

(a)

(b)

Rotated section of plan

Development

4.17 *Intersection of a plane with a cylinder with its development.*

The thicker lines show another plane intersecting the tetrahedron asymmetrically. This plane, PQ, is at 45° to the base, as shown in the front view. The line RS is horizontal.

The points P and Q can be projected directly into the plan and the side elevation. If the point R is projected into the section XX, the location of S on the rear edge CD can be found and projected back into the plan. Alternatively, the point S can be found in the side elevation (on the same level as R) and the dimension Y can be transferred to the plan.

(b) Intersection of a plane with a cylinder

In Fig. 4.17 the two front elevations show that the end of the cylinder has been cut at an angle, i.e., it has been intersected by a plane.

The plan view of the cut end is built up by taking a number of sections and showing them in both views and also in the end elevation.

In Fig. 4.17 (a) the point P is located any-

where in the elevation, and is projected into the plan and the end view. The dimension X must be transferred to the plan from the end view.

The symmetrical location of sections shown in Fig. 4.17 (b) is better. The end elevation and a rotated section of the plan are divided radially (say at 30° intervals). These section lines are carefully referenced. In this example, points 1 and 7 lie at the top and bottom of the cylinder respectively. You should project lines into the elevation to locate points systematically on the intersecting plane, and label them. Then, create verticals from these points into the plan to intersect horizontals from the rotated section. Points in the plan are located at the intersection of lines with the same reference.

The development of the cut cylinder appears almost as a byproduct of this exercise. Again, the arc is approximated to the chord between adjacent numbers on the periphery of the

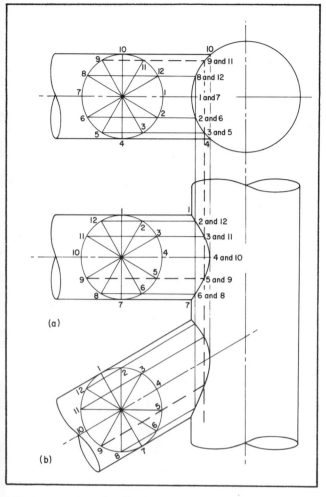

4.18 *Intersection of cylinders.*

cylinder.

(c) Intersection of cylinders

In the arrangement shown in Fig. 4.18, the plan view can be completed at once, but the elevation must be built up. In (a) the branch cylinder is normal to the main one, whereas it is inclined, or oblique, in (b).

Again a symmetrical radial division of the intersecting cylinder is used in plan and elevation with each point being given the same reference in each view. Note that rotated sections are used instead of end views for the radial division and reference numbers.

Now project lines from any reference, say 9, into both views (heavy dotted lines). From the intersection of one of these lines with the intersecting surface in the plan, drop a vertical to intersect the other line in the elevation to establish the point 9 on the intersection.

In Fig. 4.19 a variation of the same theme is shown, the intersection of a straight cylinder with a curved one.

Neither plan nor elevation can be completed without some construction.

(a) Draw an auxiliary plan of the smaller cylinder intersecting a vertical section of the larger one. Use the symmetrical radial division of rotated sections in all three views, plan, elevation, and auxiliary plan, and number similar points.

(b) Project horizontals from the same reference 3, in all three views.

(c) From the intersection of one of these horizontals with the cutting face (the larger cylinder) in the auxiliary plan, erect a vertical to meet the horizontal OH at 3′ in the elevation. OH is the start of the bend.

(d) Swing an arc, radius 03′, to cut the horizontal from reference 3 in the elevation and locate the point on the intersection.

section.

(e) Erect a vertical from this point in the elevation to intersect the third horizontal in the plan to locate point 3 therein.

4.19 *Intersection of a straight cylinder with a curved one.*

Auxiliary plan

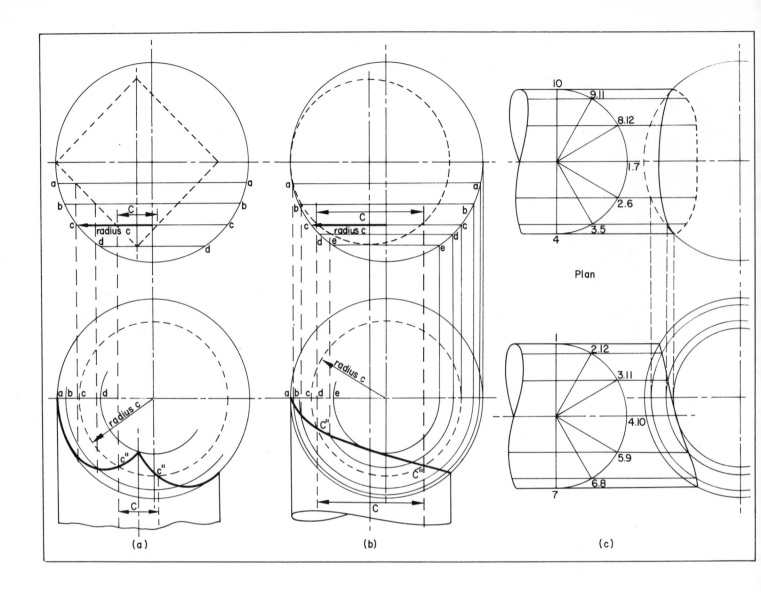

Plan

(a)   (b)   (c)

4.20 *Intersection of a sphere with plane and curved surfaces.*

(d) Intersection of planes and cylinders with a sphere

Sometimes the intersection of a doubly curved surface with a plane or other surfaces can be surprisingly simple. This is illustrated by Fig. 4.20 (a) which depicts a sphere sitting on a square base and Fig. 4.20 (b) which depicts a sphere sitting on a cylindrical base.

As in the previous examples, the construction can best be understood if sections are taken through both the intersecting shapes on the same plane. In examples (a) and (b), a vertical section CC, as defined on the plan, is taken.

At section CC, the sphere has a radius c and the base tube is C wide. The section shown in heavy dotted lines in the elevation would result if the assembly were to be cut on CC. The two points, C″, where the sections of the sphere and the base tubes intersect, lie

on the intersection being constructed.

The construction points on the intersections are fairly evenly spaced. While, in the plan, cross sections are crammed together towards the outer limits of the cylindrical base. Do not start constructions with too many points. In this case, four sections were used initially for both examples. This proved to be enough for the square base. Extra points were provided for the cylindrical base. Their location was determined by the requirements of the construction of the intersection in the elevation.

In (c) the sections are shown related to the familiar symmetrical radial division of the intersecting cylinder in both the plan and elevation. The same references apply.

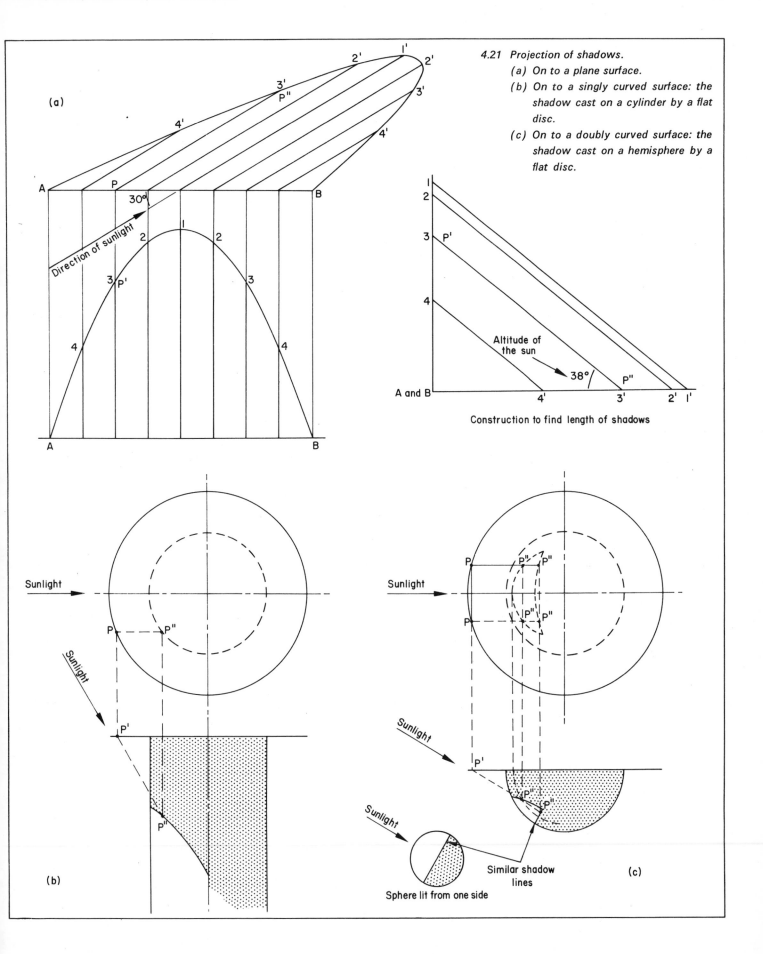

4.21 Projection of shadows.
(a) On to a plane surface.
(b) On to a singly curved surface: the shadow cast on a cylinder by a flat disc.
(c) On to a doubly curved surface: the shadow cast on a hemisphere by a flat disc.

Construction to find length of shadows

Sphere lit from one side

4.6 SHADOWS

Although engineers do not usually show shadows on their drawings to give depth or roundness to elevations, a study of the projection of shadows is a useful extension of three-dimensional thinking and some help in visualization.

If sunlight projects a shadow at right angles on to a plane surface, the shadow is an orthographic silhouette of the elevation presented to the plane, like the silhouette 'portraits' which were popular in the nineteenth century. Usually however the rays of light fall obliquely on to a plane or even a curved surface with consequent distortion of the shadow. Our own long shadows, cast by the setting sun, are a familiar example of this.

Three simple examples of shadow projections appear in Fig. 4.21 where shadows are cast on both flat and curved surfaces.

The three-dimensional problem involving the horizontal and vertical angles of the rays of the sun can be reduced to two dimensions. This is done by basing the construction, for finding the length of the cast shadows, on an axis determined by the horizontal angle of the sun's rays. In Fig. 4.21 (a) where the sunlight strikes the parabolic arch asymmetrically, the length of the shadows is determined on an auxiliary plane, whereas the symmetry of the objects in (b) and (c) makes it possible for the construction to be made directly in the elevation.

In each of these three cases, a point P is chosen on the edge of the shape casting the shadow, which is represented by the point P' in the elevation where the vertical angle of the sun is taken into account. The comparable point on the shadow is P''. The visualization of (c) is helped if you consider the appearance of the moon at the end of its first or third quarter.

4.7 EXERCISES

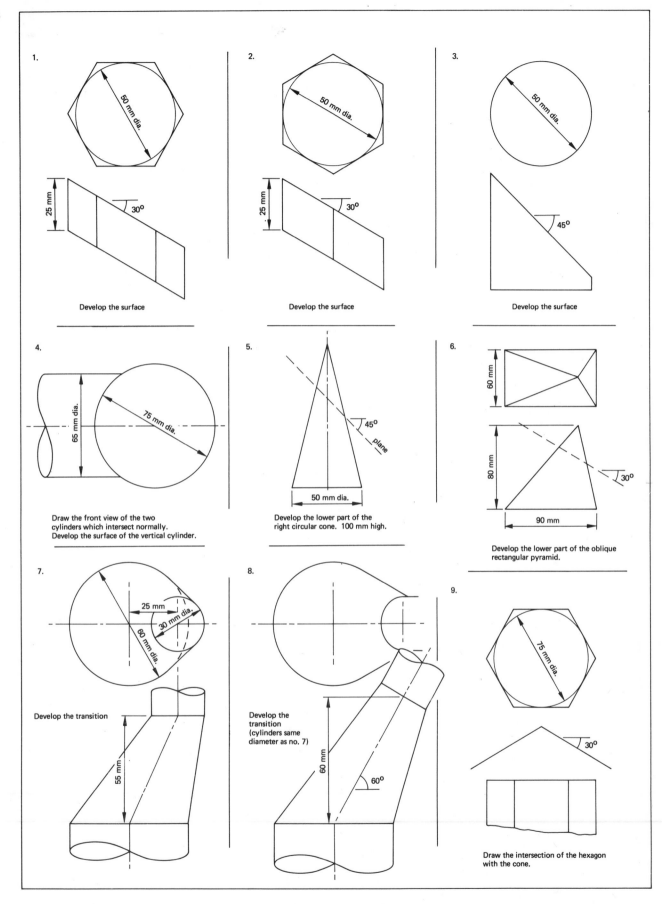

1.
50 mm dia.
25 mm
30°
Develop the surface

2.
50 mm dia.
25 mm
30°
Develop the surface

3.
50 mm dia.
45°
Develop the surface

4.
65 mm dia.
75 mm dia.
Draw the front view of the two cylinders which intersect normally. Develop the surface of the vertical cylinder.

5.
45°
plane
50 mm dia.
Develop the lower part of the right circular cone. 100 mm high.

6.
60 mm
80 mm
90 mm
30°
Develop the lower part of the oblique rectangular pyramid.

7.
25 mm
30 mm dia.
60 mm dia.
55 mm
Develop the transition

8.
Develop the transition (cylinders same diameter as no. 7)
60 mm
60°

9.
75 mm dia.
30°
Draw the intersection of the hexagon with the cone.

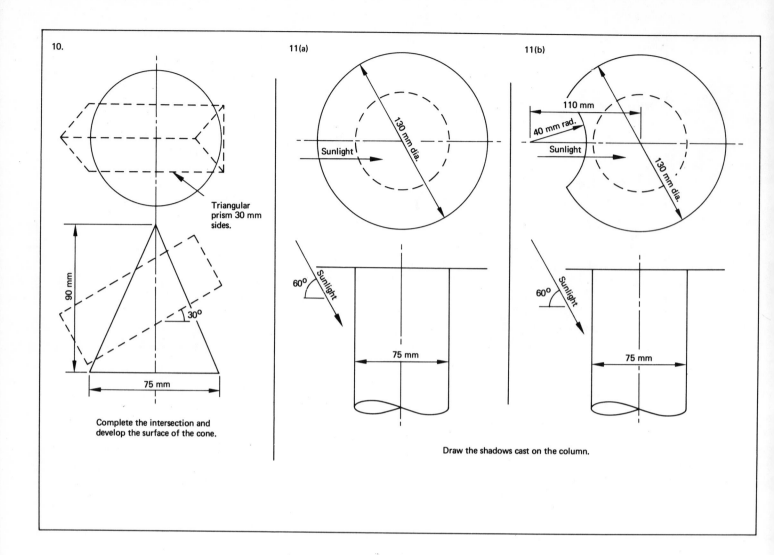

10.

Triangular prism 30 mm sides.

90 mm

30°

75 mm

Complete the intersection and develop the surface of the cone.

11(a)

130 mm dia.

Sunlight

60° Sunlight

75 mm

11(b)

110 mm

40 mm rad.

Sunlight

130 mm dia.

60° Sunlight

75 mm

Draw the shadows cast on the column.

Chapter 5   **Conics**

Road transport, Scotswood Bridge and approaches. A neat road bridge, of welded steel tied-arch construction with its deck suspended from the arch, together with the complicated approaches necessary to maintain traffic flow at peak hours. (Photograph supplied by Mitchell Construction, Kinnear Moodie Group)

5.1 INTRODUCTION

A conic is a curve resulting from the inter-section of a plane with a cone. The civil engineer makes considerable use of these curves, either because they are pleasing shapes that are also structurally functional, or because they appear in his calculations.

The graphical derivation of the conic is a useful exercise, for the engineering student, in three-dimensional thinking and auxiliary views. The derived curves have useful applications in practice.

5.2 DERIVATION

The four fundamental conics, the circle, the ellipse, the parabola, and the hyperbola, may be obtained as the sections of a right circular cone, as shown in Fig. 5.1.

The axis of a right circular cone passes through the centre of the circular base and is normal to it.

(a) The circle is a section on a plane nor-mal to the axis of the cone.

(b) The ellipse is a section on a plane, other than the one described in (a), that cuts right through the cone.

(c) The hyperbola is a section on a plane parallel to the axis of the cone. Note that this plane can cut another cone balanced symmetrically on top of the first cone.

(d) The parabola is a section on a plane between those in (b) and (c). In these exercises, it is taken parallel to an ele-ment, i.e., parallel to the sloping edge of an elevation.

5.3 CONSTRUCTION OF A CONIC FROM A GIVEN CONE

Figure 5.2 shows a plane, an, cutting through a right circular cone to produce a parabola.

The method described can also be used for an ellipse or hyperbola. It consists in taking horizontal sections of the cone.

(a) Set off equal increments along the plane, an, in the elevation, ab, bc, etc.

5.2 *The general geometric construction of a conic section applied to the parabola.*

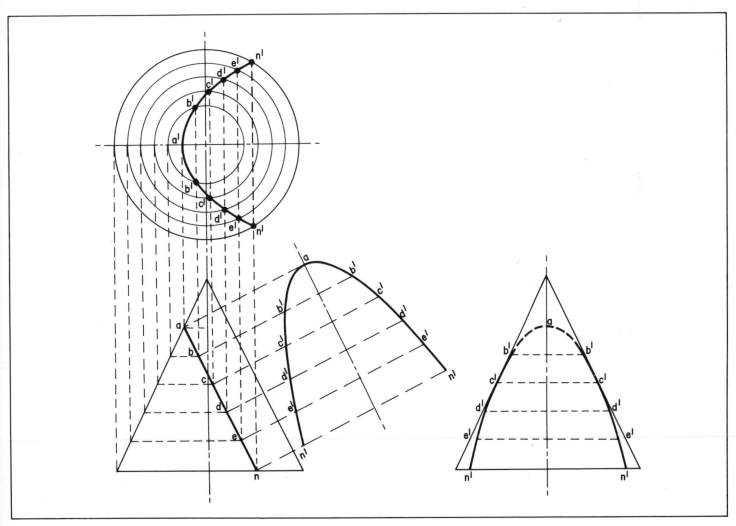

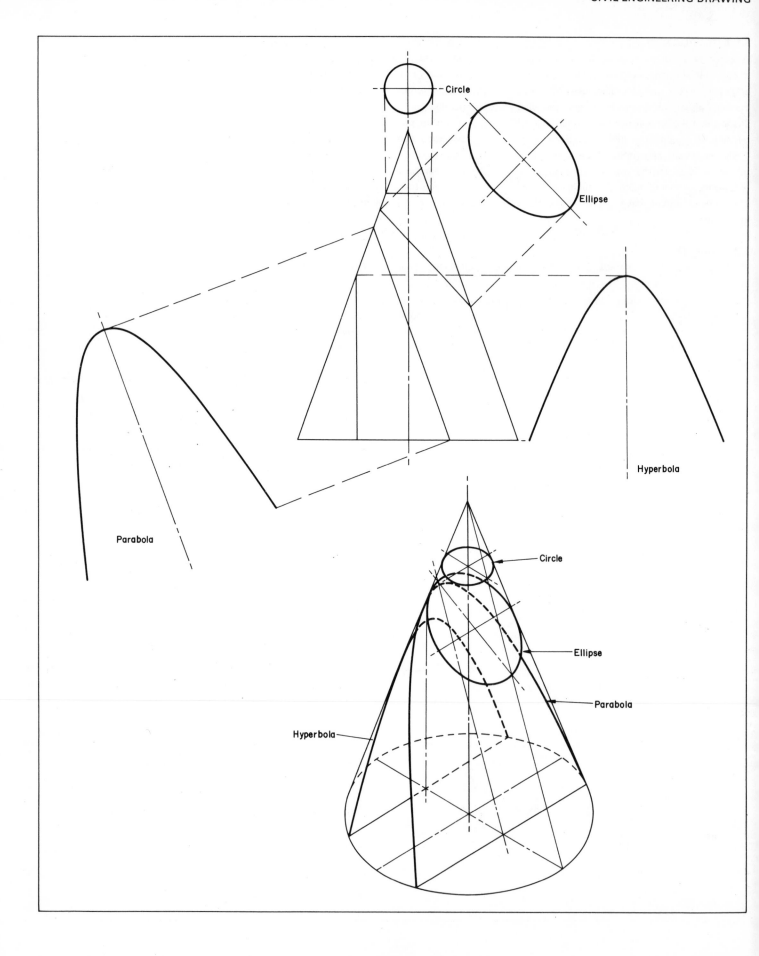

Circle

Ellipse

Hyperbola

Parabola

Circle

Ellipse

Parabola

Hyperbola

Note: an will have a finite length for the ellipse and should be divided into equal increments so that the construction is symmetrical. It is difficult to draw a symmetrical shape through points arranged asymmetrically.

(b) Take sections of the cone at a, b, c, etc., which will appear in the plan as circles.

(c) The intersecting plane will cut the circles in the plan at a', b'b', c'c', etc., vertically above the points a, b, and c in the elevation. The lines b'b', c'c', etc., will be the true widths of the parabola at a, b, c, etc., and can be plotted on the side elevation where a foreshortened view of the parabola will appear. An auxiliary view will be necessary to show the true shape of the parabola.

5.4 FIVE METHODS OF GRAPHICAL CONSTRUCTION OF AN ELLIPSE

The ellipse was probably used more extensively in the nineteenth century when masonry and brick arch bridges were being constructed. The ellipse replaced the circular arches as it allowed long spans to be made without excessive rise. The arch Brunnel put over the River Thames at Maidenhead was so flat, it is said that he was made to leave the timber formwork in place in case the arch fell down. He is reputed to have eased the timber before he left the bridge. In any case, the bridge still stands and carries rail loads unthought of when it was built.

(a) Strings and pegs

Figure 5.3 shows how an ellipse may be generated by a point moving so that the sum of its distance from two points, the foci, is constant and equal to the major axis. This method is little used on the drawing board, but is used out of doors, often by groundsmen.

The foci are located by swinging arcs, with a radius a equal to half the major axis, from the ends of the minor axis C or D, to cut the major axis at the points F.

Set up pegs at the foci. Make a loop with a piece of cord of length (AB + FF), i.e., the major axis plus the separation of the foci. Keep the cord tight round the pegs; the point P will then trace out the required ellipse.

(b) Trammel method

This is a useful method in practice as points can be located at will to define the tighter parts of the curve.

The trammel, see Fig. 5.4, can be of folded paper or card. Set off the semimajor axis a and the semiminor axis b in either of the two ways shown, and mark the points A, B, and P.

Move the trammel so that B moves up and down the minor axis while A moves along the major axis. The point P traces out the ellipse.

(c) Concentric circle method

Draw a pair of concentric circles with radii equal to half the two axes of the ellipse, see Fig. 5.5.

From the centre of the circles, draw radial lines to cut the circles at A and B. A point P on the ellipse is located at the intersection of a vertical from A and a horizontal from B.

Locate a small number of points P to start with and provide extra ones where needed to define the tighter parts of the curve.

(d) Approximate arcs method

An ellipse may be approximated by two

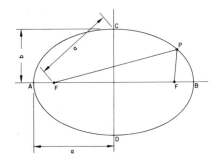

5.3 *Graphical construction of an ellipse by string and pegs.*

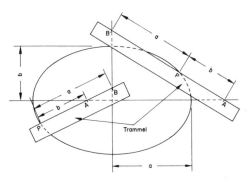

5.4 *Graphical construction of an ellipse by the trammel method.*

5.5 *Graphical construction of an ellipse by concentric circles.*

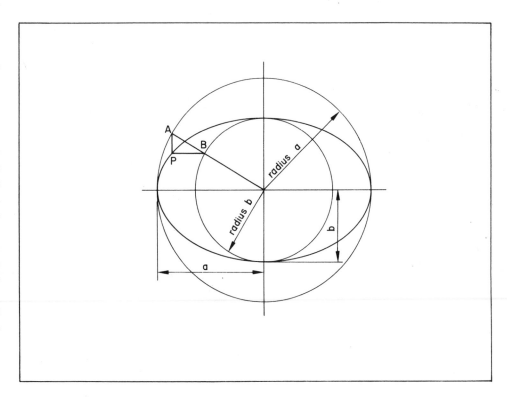

5.1 *The four sections of a cone.*

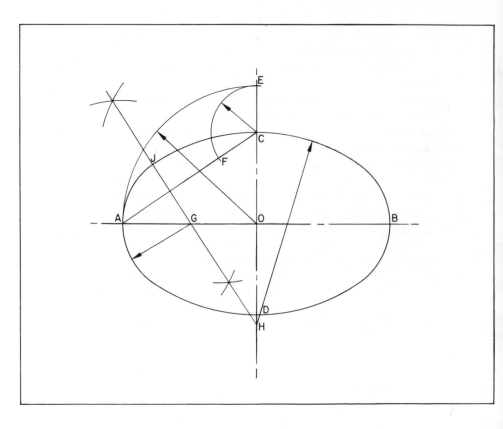

5.6   *Graphical construction of an ellipse by approximate arcs.*

pairs of circular arcs of different radius, see Fig. 5.6, using the following method.

(a) Set up the two axes.

(b) Join AC.

(c) Centre O, swing an arc, radius OA, to cut OC produced at E.

(d) Centre C, swing an arc, radius CE, to cut CA at F.

(e) Locate points G and H where the perpendicular bisector of AF cuts OA and OD, produced if necessary.

(f) Centre H, swing an arc, radius HC, to complete the part of the ellipse CJ.

(g) Centre G, swing an arc, radius GA, to J.

It is useful, when sketching, to draw an ellipse as two pairs of circular arcs; the result is very realistic, and more symmetrical than would otherwise be achieved.

5.5 TWO METHODS OF GRAPHICAL CONSTRUCTION OF A PARABOLA

The parabola is perhaps the curve most used in civil engineering. You will be well advised to remember at least one quick graphical method of construction, probably method (a) described below.

A French curve, with a good range of parabolae, is essential for good results. Many French curves are too small and their curves too sharp for constructing most parabolae.

In practice, you may find that the greatest use of parabolae will be in constructing bending moment diagrams for distributed loads. Time can be saved as shown in Fig. 5.7, if the points of maxima and minima moment are found and the parabolae between them constructed graphically rather than by calculating the moments at a number of points, plotting them, and then drawing the curve.

Find the value of the bending moment at as many points as possible, in this example A, B, C, and D. The maximum positive moment will occur where the shear force is a minimum. The parabolae AB' and BC' may then be constructed.

Note that the area of a parabola is 2/3 the area of the enveloping rectangle.

The parabola is sometimes considered to be more pleasing aesthetically than the circular arc. Even when the parabola is very flat, for example, at the springings of a flat arch, the subtle difference is more satisfying to the eye than a continuous arc. From a practical aspect, the slope at the outer ends of a parabola is flatter than at the same points of a circular arch with the same span and rise. This often makes the parabola more convenient for bridge deck profiles. Few things are more pleasing visually than the sweep of the cables of a suspension bridge and these are very

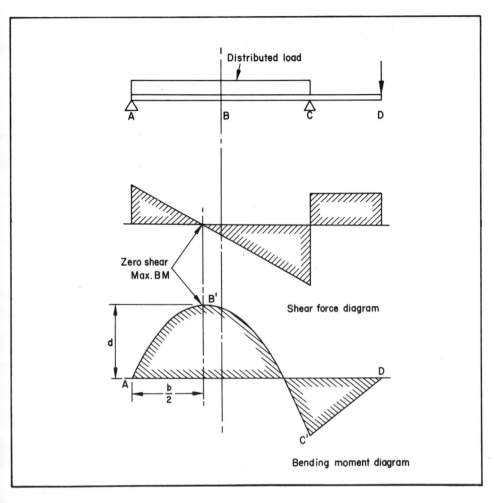

Distributed load

Zero shear
Max. BM

Shear force diagram

Bending moment diagram

nearly parabolic.

(a) Graphical construction to a given rise and span

To draw a parabola of span b and height d, as shown in Fig. 5.8 (a), the following instructions should be remembered. This is the most likely way a parabola will be specified.

(a) Divide OA into any number of equal parts and number them as shown. Drop vertical lines from the points 1, 2, 3, etc.

(b) Divide AB into the same number of equal parts and number them as shown.

(c) Points P on the parabola will occur at the intersection of the sloping lines O1, O2, O3, etc., with the verticals through 1, 2, 3, etc.

(b) Graphical construction between two given tangents

A parabola, drawn between two given tangents, see Fig. 5.8 (b), is useful for sketching vertical road curves quickly so that alternative schemes can be appraised. Levels for con-

struction of the chosen scheme would then be calculated. The method is also applicable to skewed parabolae which might be needed for aesthetic reasons.

(a) Produce the tangents to intersect at A. The parabola meets the tangent at B and C.

(b) Divide AC and AB into the same number of equal parts.

(c) Number one tangent from the tangent point C to the intersection point A, and the other one in the reverse direction, A to B.

(d) Join points with the same number, 1 and 1, 2 and 2, 3 and 3, etc., to form a series of additional tangents to the parabola.

(e) Draw the parabola as a smooth curve within the envelope of tangents.

5.6 GRAPHICAL CONSTRUCTION OF AN HYPERBOLA

Figure 5.9 shows how to draw an hyperbola given the asymptotes OA, OB and a single point P.

5.7 *Uniformly distributed loading on a beam giving rise to a parabolic bending moment diagram.*

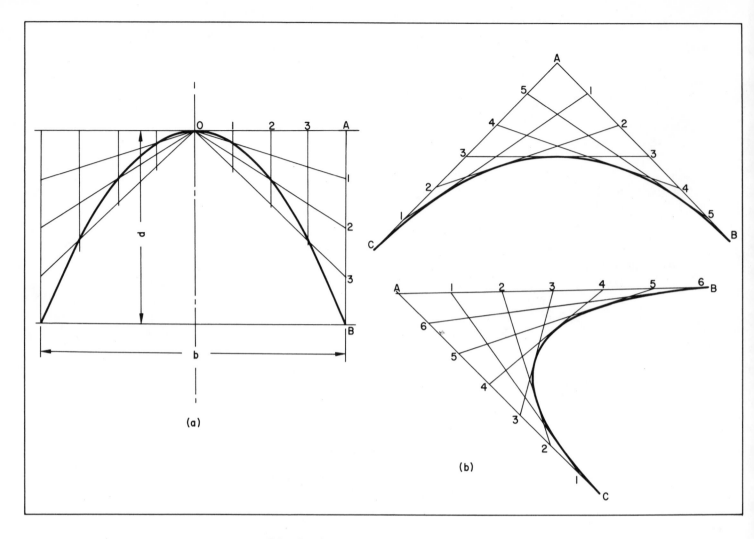

(a)

(b)

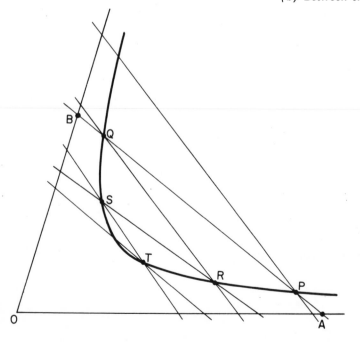

5.8   *Graphical construction of a parabola.*
(a) To a given rise and span.
(b) Between two given tangents.

The construction is based on the fact that intercepts AP, BQ, on a chord of the hyperbola, are equal.

(a) Draw any chord across the axes or asymptotes to pass through point P.

(b) Measure intercept AP and set off BQ equal to AP.

(c) Draw any further chords through P or Q and establish additional points R, S, T, etc., as necessary.

5.9   *Graphical construction of an hyperbola given the assymptotes AO, AB and a single point P.*

5.7 EXERCISES

1. The ellipse

   Compare the three methods of drawing an ellipse, trammel, concentric circles, and approximate arcs, by constructing an ellipse, 150 x 100 mm, by each method on tracing paper and laying the ellipses over each other. Then construct ellipses with various sizes of major axis to see where the differences are more marked.

2. The parabola

   (a) Draw two parabolae by the method given in section 5.5 (a):

   (i) 60 mm rise by 180 mm span.
   (ii) 150 mm rise by 75 mm span.

   (b) Draw the bending moment diagram for a beam 30 m long, supported at one end and at 6 m from the other end, with a total load of 100 Kg/m over its whole length.

3. The hyperbola

   (a) Complete a rectangular hyperbola between the x and y axes if it passes through the point (4.1).

   (b) Draw an hyperbola between asymptotes 60° apart which passes through a bisector of the angle 30 mm from the intersection.

4. Draw an elevation of an arch 20 m rise by 60 m span and compare the three curves:

   (a) Circular arc.
   (b) Ellipse.
   (c) Parabola.

Chapter 6   **Measured perspective**

High Rise, a development at Tolworth, Surrey, consisting of a shopping centre, offices, and parking area. (Photograph supplied by the Cement and Concrete Association)

6.1 INTRODUCTION

Although small objects can be shown satisfactorily in isometric projection, the lack of perspective makes it unsuitable for larger objects such as buildings, dams, etc. The human eye is used to seeing parallel lines converge as they run to infinity and is disturbed if they do not.

6.2 SINGLE POINT PERSPECTIVE

The simplest perspective drawing is single point perspective in which all parallels converge at one point. Single point perspective results if the centre of vision (see definitions below) is parallel with one axis of the scene, e.g., looking down a straight railway track or standing in a room facing one wall squarely and looking parallel to another.

Single point perspective is shown in Fig. 6.1 to which the following definitions apply.

(a) Station point (SP): The point from which the object or scene is viewed. It is better not to call it the view point in case the abbreviation VP is confused with the vanishing point, see (d) below.

(b) Horizon: The horizontal plane which contains the station point.

(c) Picture plane (PP): The plane on to which the perspective is projected.

(d) Vanishing point (VP): The point at which parallel lines seem to join to-

6.1 *Single point perspective: general construction showing how the size of the picture is controlled by the location of the picture plane.*

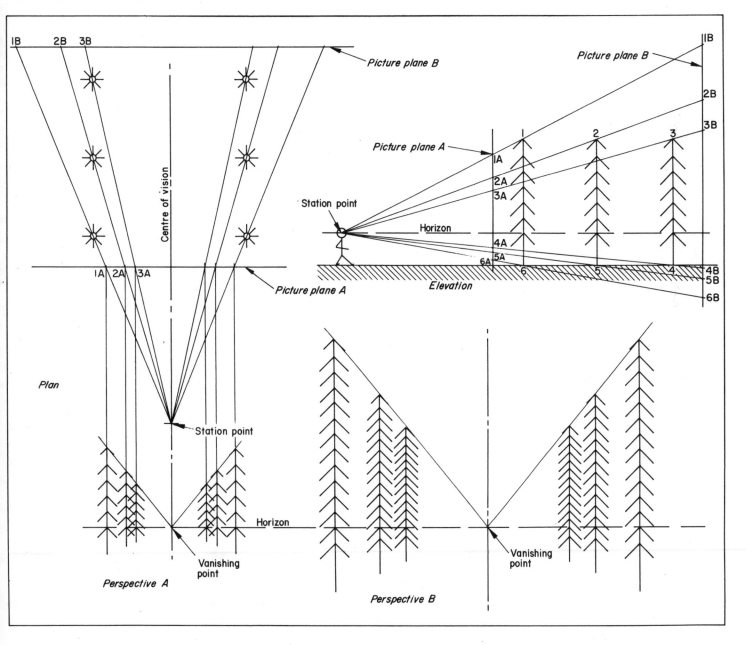

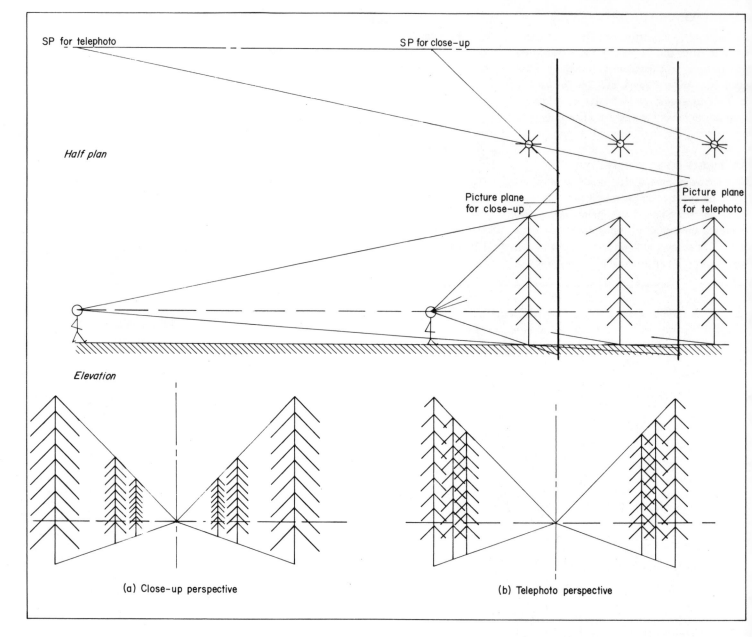

Half plan

Elevation

(a) Close-up perspective   (b) Telephoto perspective

6.2   *Single point perspective showing how
      the type of perspective is controlled by
      the location of the station point.*

gether or vanish. Left-hand and right-hand points are indicated by VPL and VPR.

(e) Centre of vision (CV): The optical axis of the construction.

The distance of the station point from the nearest point of the object or scene will have a considerable effect on the perspective. A close station point exaggerates the perspective by making the nearest tree seem much larger than the succeeding ones, see Fig. 6.2 (a). This technique is often used by advertising illustrators when they want to exaggerate some feature, for example, the interior of a small car can be made to appear cavernous. If the station point is too far away, the col-

lapsed perspective of the telephoto lens is obtained. Objects seem to be very closely packed with little loss of height from front to back of the scene, see Fig. 6.2 (b). A normal perspective effect is gained if the angle between the lines joining the eye to the extreme points of the object does not exceed 30°, either vertically or horizontally.

Natural perspectives are obtained with the horizon at eye level when standing, about 1.5 m above the ground. A lower horizon often gives a dramatic effect which architects like, while a bird's-eye view can clarify the layout of a project.

The plane on which the picture is to be projected should be located where it will give

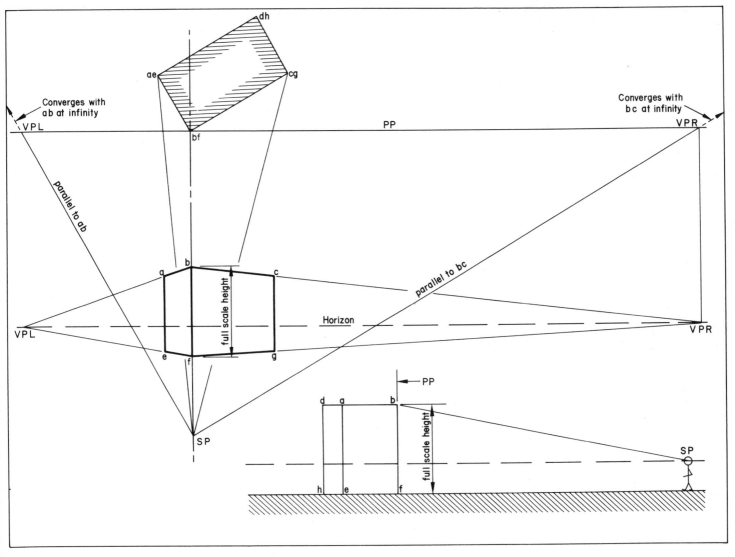

a finished picture of the required size. Picture plane B in Fig. 6.1 gives a larger picture than picture plane A without affecting the perspective. For picture plane A, imagine that a piece of glass has been placed between the observer and the scene so that a tracing can be made. Picture plane B can be considered as a screen on to which shadows of the objects are cast by a point source of light at the observer's eye. Note that the picture planes in Fig. 6.2 were located to make the nearest tree in each perspective the same height.

The centre of vision should be directed at the centre of interest. If there is no obvious centre of interest, it is usually better to split the view asymmetrically with the centre of vision, unless the symmetrical effect of Fig. 6.1 is required.

6.3 CONSTRUCTION OF SINGLE POINT PERSPECTIVE

This is shown in Fig. 6.1 and Fig. 6.2.

(a) Draw a plan and elevation to suitable scale.

(b) Decide upon and locate the horizon, station point, and picture plane.

(c) Draw lines from the station point to definitive points in the view (the tops and bottoms of the fir trees in Fig. 6.1) to cut the chosen picture plane.

The intercepts on the picture plane now represent the dimensions of the definitive points or lines. Heights appear in the elevation and widths in the plan.

It is often convenient to construct the perspective view directly below the plan, as for perspective A in Fig. 6.1, to save transferring dimensions. View B however was constructed by transferring all the dimensions.

The mechanics of single point perspective now appear; all parallels converge to a single vanishing point.

6.3 *Two point perspective: general construction showing how the vanishing points are located in the picture plane.*

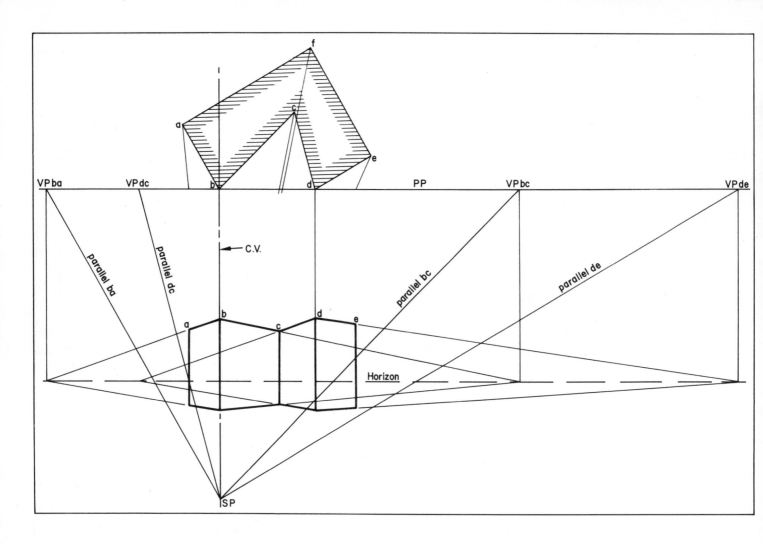

6.4   *Multi-point perspective which results if the faces of the object lie on several planes.*

6.4 TWO POINT PERSPECTIVE

If the centre of vision is not parallel with one of the axes of the object, parallels in the perspective will converge on two or more vanishing points, see Fig. 6.3 and Fig. 6.4 respectively.

Two points of note arise in Fig. 6.3.

(a) The vanishing points are on the horizon at infinity. Thus the line of sight from the station point to the vanishing point for the side of the block abfe will be parallel to that side.

(b) For simple perspective exercises, an elevation is usually not required if the picture plane passes through the object. In Fig. 6.3, the corner of the block bf lies in the picture plane and will appear in the perspective at its full height (according to the scale of the plan).

There are four vanishing points in Fig. 6.4 because the four visible faces of the block lie on different axes.

6.5 CONSTRUCTION OF TWO POINT PERSPECTIVE

Refer to Fig. 6.5.

(a) Draw a plan view. A good impression is obtained if the axes of the block are at 30° and 60° to the picture plane.

(b) Make the nearest corner the centre of interest and draw the centre of vision through it.

(c) Locate the station point so that the maximum included angle is 30°.

(d) Draw the picture plane to give the perspective view of the required size and then locate the vanishing points by drawing lines through the station point parallel to the axes of the block to cut the picture plane.

Note: At this stage, it often happens that the vanishing points are off the paper. So, for routine exercises, it is sensible to work backwards by locating the picture plane first, putting the

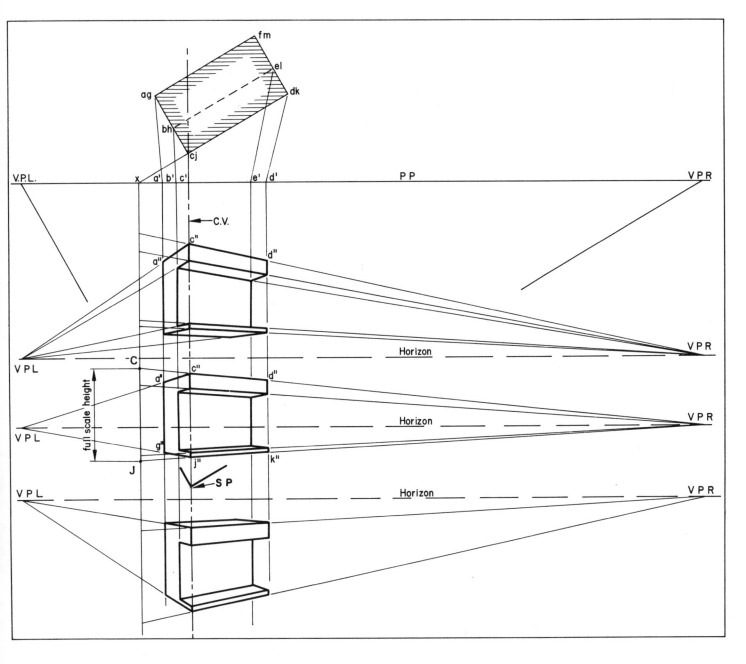

vanishing points at each end of it, and then locating the station point by dropping lines from the vanishing points parallel to the axes of the object. Finally, erect the centre of vision from the station point and draw the plan of the block.

(e) Draw lines from definitive points on the object abcde towards the station point to cut the picture plane at a′ b′ c′ d′ e′. It is better not to continue the lines to the station point as the mass of radiating lines can obscure subsequent construction.

(f) Start the perspective view by drawing its horizon and erect the first vertical.

Note: In the case of Fig. 6.3 the full scale height of the first vertical was used. In this case, the picture plane does not cut the object, therefore,

(i) project dc to meet the picture plane at X where cj will appear to its full scale height;

(ii) drop a vertical from X into the perspective view and set up CJ to its full scale height;

(iii) join C and J to the vanishing point right when the required

6.5 *Two point perspective: detailed construction showing the effect of raising or lowering the horizon beyond the vertical limits of the object.*

height c″ j″ will be the intercept on the vertical from c′, which, in this case, is also the centre of vision.

(g) Find d″ vertically below d′ and on the line from c″ to the vanishing point right.

Bird's-eye or worm's-eye impressions can be drawn by moving the horizon beyond the lower or upper limits of the object.

A distorted effect will be obtained if the horizon is removed too far from the vertical limits of the object because the verticals should then converge to vanishing points, see the three point perspective in Fig. 6.6 which is not pursued here.

Two point perspective can be adopted, as shown in Fig. 6.7, to give vertical perspective if the construction is based on an elevation instead of a plan. The centre of vision can be moved, like the horizon in the previous examples, to reveal the side of the object.

6.6 REALISTIC COMPLETION OF PERSPECTIVES

Any draughtsman can perform the mechanics

of perspective drawing and end up with a characterless skeleton, only the gifted can bring perspectives to life, as shown in Fig. 6.8 (b). An engineer with a knowledge of perspective will at least be able to argue intelligently with the artist or architect who is to make the artist's impression of his project. The engineer might prevent too outrageous an interpretation being made, or provide the artist with an accurate skeleton to clothe.

In any case, an engineer should keep a sketch book of his ideas or ideas he has borrowed from other projects. A knowledge of perspective will speed and improve his sketches.

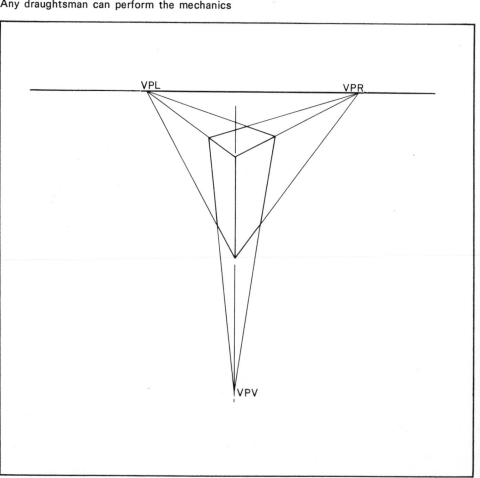

6.6   Three point perspective: general construction.

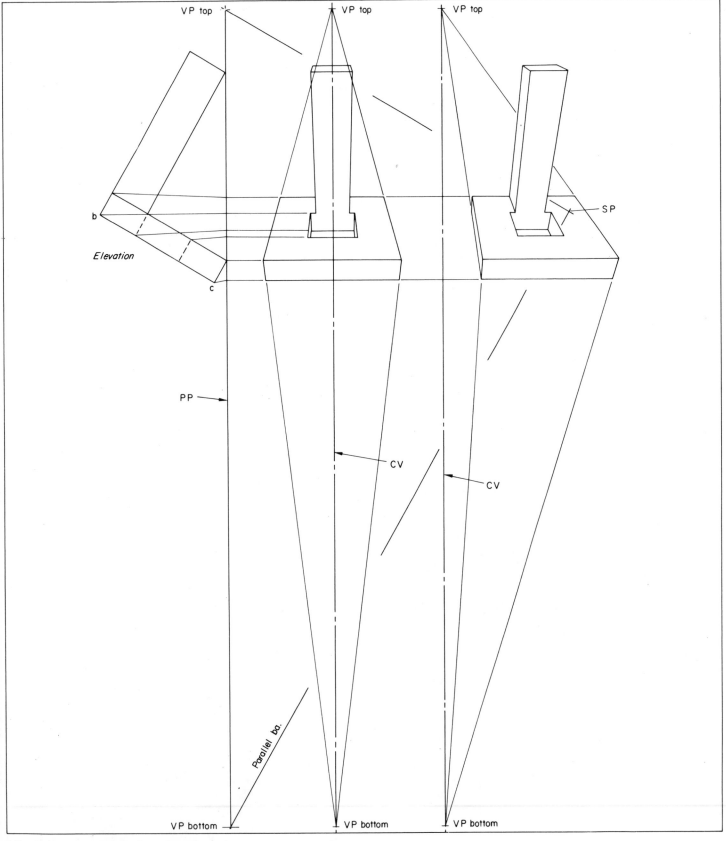

6.7 *Two point perspective adapted to give vertical perspective to a tall object. The construction is based upon an elevation instead of a plan.*

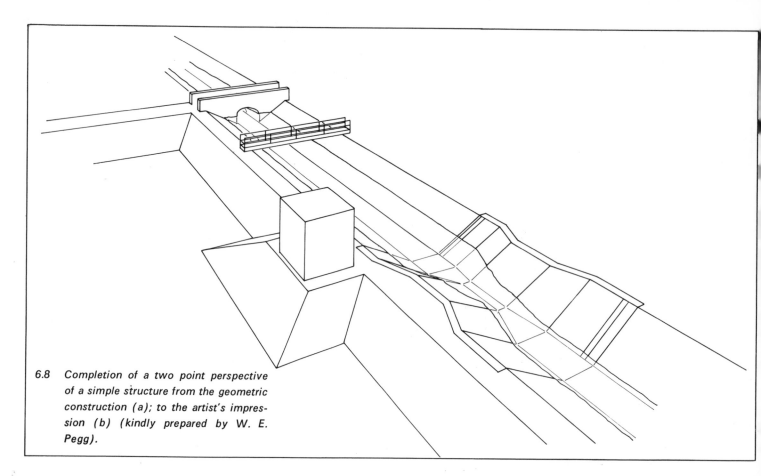

6.8   *Completion of a two point perspective of a simple structure from the geometric construction (a); to the artist's impression (b) (kindly prepared by W. E. Pegg).*

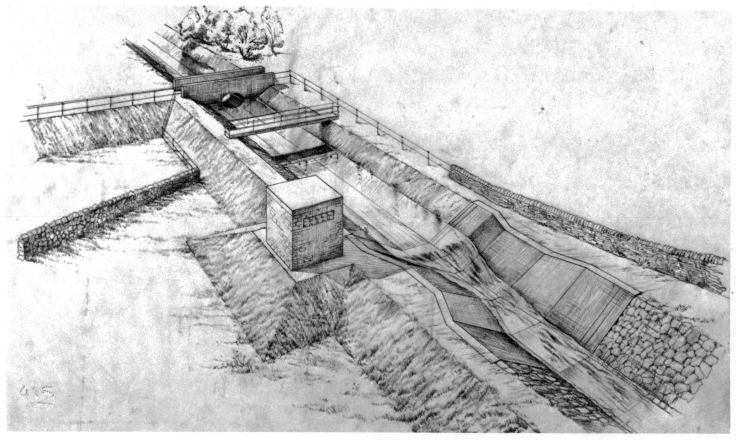

6.7 EXERCISES

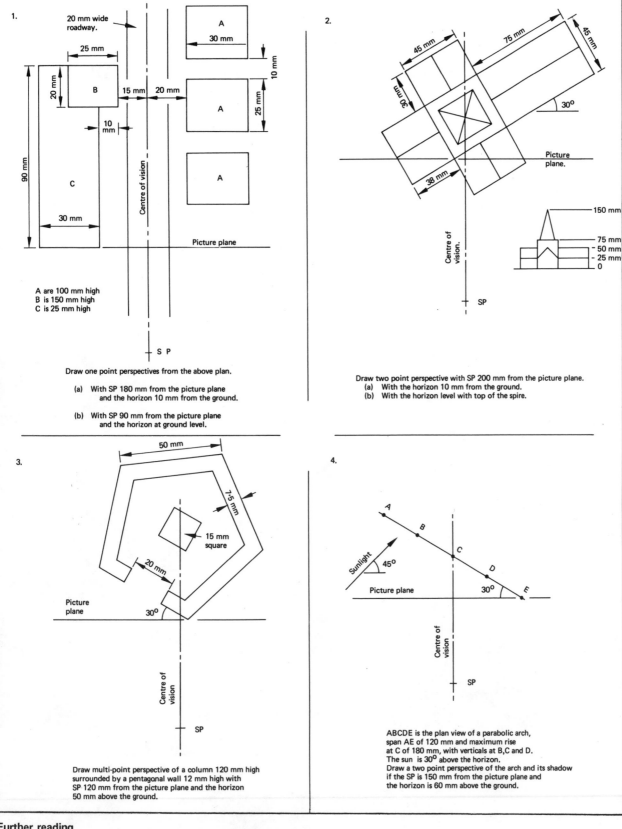

1.

A are 100 mm high
B is 150 mm high
C is 25 mm high

Draw one point perspectives from the above plan.

(a) With SP 180 mm from the picture plane
      and the horizon 10 mm from the ground.

(b) With SP 90 mm from the picture plane
      and the horizon at ground level.

2.

Draw two point perspective with SP 200 mm from the picture plane.
(a) With the horizon 10 mm from the ground.
(b) With the horizon level with top of the spire.

3.

Draw multi-point perspective of a column 120 mm high
surrounded by a pentagonal wall 12 mm high with
SP 120 mm from the picture plane and the horizon
50 mm above the ground.

4.

ABCDE is the plan view of a parabolic arch,
span AE of 120 mm and maximum rise
at C of 180 mm, with verticals at B,C and D.
The sun is 30° above the horizon.
Draw a two point perspective of the arch and its shadow
if the SP is 150 mm from the picture plane and
the horizon is 60 mm above the ground.

Further reading

1. Capelle:

Professional Perspective Drawing for Architects and Engineers, McGraw-Hill Book Company Inc.

2. Martin:

Design Graphics, Collier- Macmillan.

Part Three In practice

Chapter 7 Drawing in practice—a case history

Modern hyperbolic roof structure in a natural setting. The Commonwealth Institute, Kensington, London. (Photograph supplied by the Cement and Concrete Association)

7.1 INTRODUCTION

In this chapter the development of civil engineering schemes will be studied very generally from conception to completion, indicating at each stage the type of drawings that can be involved. Many actual examples have been included.

For the purposes of the case history, it is assumed that a local authority is preparing a scheme large enough to require a special Act of Parliament for its execution. Consulting engineers might be engaged for the design and supervision of the construction, but this would not affect the type and range of drawings used.

The graphical programme is divided into simple steps relating to the general headings to each section.

7.2 FUTURE NEEDS

In addition to supervising the day to day running of his office and dealing with schemes already under construction, the engineer to any authority must be aware of future needs and make plans to meet them.

The city engineer, *inter alia*, must try to foresee future traffic congestion, the need to clear rundown residential areas, whether the sewage system and treatment works will be adequate in the future. The water engineer must be sure that, in addition to having enough water, see Fig. 7.1 (a) and (b), his treatment works are satisfactory and that his pipelines are biologically and structurally safe. The land drainage engineer must be sure that his watercourses, sluices, and locks are adequate and safe and that his pumping stations are large enough and up to date. The engineer responsible must know whether the airport runways and terminal buildings are large enough to meet future needs; whether the anchorage is deep enough for larger tankers; whether the track of the main line is still safe since the development of the nearby boggy land.

A guide to the solution of some of these problems lies in the trend of population in the area and the standard of living that will be expected within the time being considered.

If the area is a thriving one, as the south-eastern corner of England is at present, not only will the population increase, but the standard of living will tend to rise with associated increases in the *per capita* consumption of water and electricity. People will want more motor vehicles and aircraft, while they and their factories will create more sewage and waste, the latter often of increasing toxicity and with its associated problems.

Whether he is looked upon as prescient, courageous, optimistic, or a fool to try, the engineer must attempt to forecast his employers' requirements.

The need for thrift in expenditure or the spreading of funds as far as possible means that, at best, a scheme must not be completed too far in advance of needs, while the norm is often one crisis of survival after another.

Some authorities never catch up, but others are prepared to invest for the future. It is up to the engineer to try to understand his authority so that he can present to them the scheme they want or, better still, make them want the scheme he presents. In any case, well prepared, easily understood drawings will often carry the day for the engineer. A committee, not made conscious of their possible inability to read tricky engineering drawings, can be more receptive and amenable to new ideas.

7.3 SEARCHING FOR A SITE

Whether the engineer is looking for a site for a sewage works, railway siding, or reservoir, he will employ maps or aerial photographs to speed the search. Although Great Britain is covered by the most comprehensive system of mapping of any country in the world, the very best map cannot be interpreted fully without a visit, however cursory, to the site.

In addition to the wide range of topographical maps, the engineer has at his disposal maps of many diverse types, including geological, land use, population concentration, rainfall, etc.

While aerial photographs, vertical or oblique and preferably stereoscopic, are not generally available, every effort should be made to obtain them. One of the firms that specializes in this work may already have covered the area for some other purpose. No scheme of any size should be planned without aerial photographs. So many engineers sadly relate how a fault or hazard, that upset their plans and wasted time and money, was obvious when they subsequently examined aerial photographs. Complete interpretation of aerial photographs is a job for the specialist.

Some schemes, particularly in underdeveloped countries, often rely entirely on photographs and photogrammetry of the type shown in Fig. 8.17 (a) for site information. Where no reliable or up to date maps exist, aerial photographs backed up by ground control and speeded up by electronic measuring devices, are the only way of obtaining site information in a reasonable time and at reasonable cost. In our own relatively overdeveloped land, some local authorities use aerial photographs to find developments which have not been approved.

We are not concerned here with the social and political factors which influence the engineer when he is looking for a site, but they can be an even greater hazard than nature itself. The ponderous machinery of land acquisition, whether achieved voluntarily or compulsorily, and the time involved thereby can often influence the choice of site and result in a less suitable one being selected. Over half the time taken in completing a motorway can be consumed in the litigation and archaic processes of land acquisition. The engineer is not a tyrant needlessly dispossessing people of their houses and land. However, he must be prepared for many frustrations and delays to his plans while land agents and lawyers carry on their protracted negotiations.

The plans used to acquire land are important documents in this phase of the scheme. They are usually based on suitable *Ordnance Survey* sheets, often 1:2500 scale, with the land to be acquired indicated in some way, see Fig. 7.2. Usually, the land to be acquired is coloured red. Different colours are generally used for land required for working space and construction access or for land over which access may occasionally be required in the future, a wayleave. The preparation of these plans can consume much drawing office time. However, transparent photocopies of the appropriate *Ordnance Survey* sheet for making dyeline prints, are being used more and more to speed this work.

7.4 DESK STUDIES

When several suitable sites for a project present themselves, their topographical suitability can often be compared, with sufficient accuracy, from desk studies based upon site information taken from maps and other available sources. The feasibility studies for the Morecambe Bay and Solway barrages were carried out in this way.

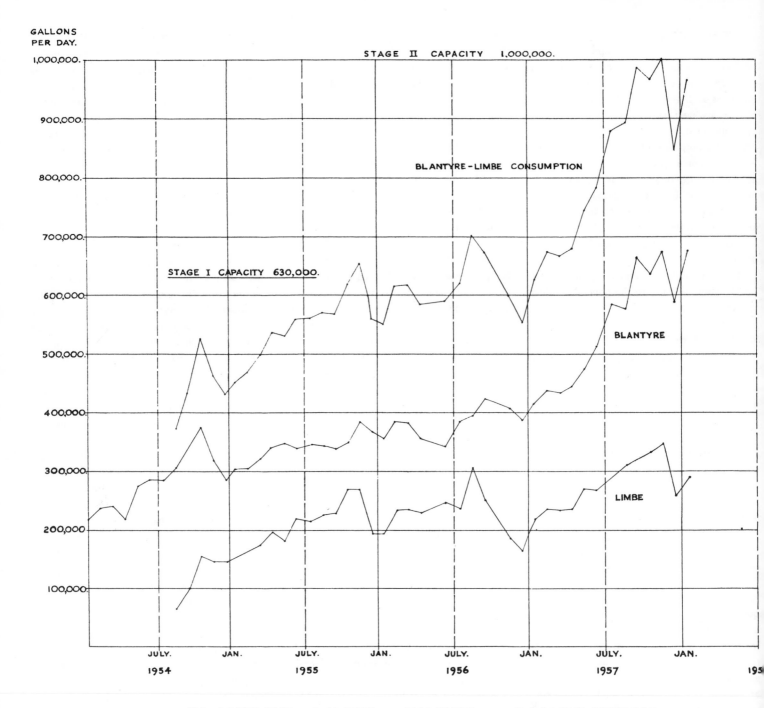

BLANTYRE - LIMBE   WATER   CONSUMPTION

7.1 (a) *Record of water consumption for two African towns, showing the growth in consumption.*
(Supplied by Scott & Wilson, Kirkpatrick & Partners)

If, for example, a reservoir is to be sited in a given valley which has reasonably predictable geology, the preliminary calculations of reservoir depths and capacity can be based upon the contours shown on a map of suitable scale. The dam can be moved and further studies of depth/capacity relationships can be carried out, without moving from the desk and without stirring up local feeling and causing unnecessary alarm to residents before a scheme is known to be feasible.

A development of the desk study is one from a car window. With maps of the correct scale, vast areas can be covered in a few days. For example, the preliminary alignment of a water main nearly two hundred miles long was outlined in two or three days by this method and was surprisingly close to the finished design which emerged several months later. Some local survey work had to be carried out where relative levels over flat slopes could not be determined by eye.

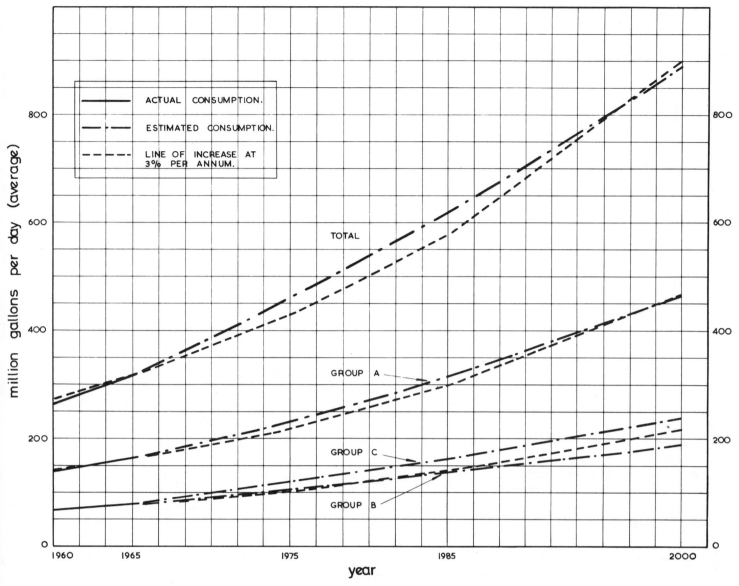

Sometimes it may be necessary to amplify the information obtained from the map with limited site surveys and investigations of subsoil and geological conditions.

Often the design process is easier in difficult terrain than in more amenable areas. A mountain road may have only one route which is physically possible, but a road across easy country can have many variants, economic, political, or social, all of which must be studied. Moreover, engineering in difficult terrain is more interesting.

Here is an extreme example of this type of study. The alignment, survey, and tender drawings for a seven mile stretch of mountain road were completed in three winter months by using the 1:2500 *Ordnance Survey* sheets as background. The route was first sketched on to the maps by eye, on site. The curves

and geometry were then adjusted to conform to vehicular speed requirements. Finally, cross sections were taken for the combined process of designing vertical curves and balancing the cut and fill of earthworks as far as possible. The intersection points of the straights, the basis for setting out the curves, were located on site by measurements from such topographical detail as existed, corners of stone walls, gateways, etc., see Fig. 7.3. Only then were the angles between straights measured by theodolite so that the curves and transitions could be ranged in. By delaying and reducing the quantity of time consuming detail, survey, and drawing, a tight contractural deadline was met, although much more responsibility was thrown upon the site supervisory staff. Originally, it was intended to use 1:500 photographic enlargements of the 1:2500

(b) Trend curve of water supply for the basin of the River Trent recording demands to date and forecasting future needs in three subdistricts as well as in the basin as a whole. (Supplied by the Engineer to the Trent River Authority)

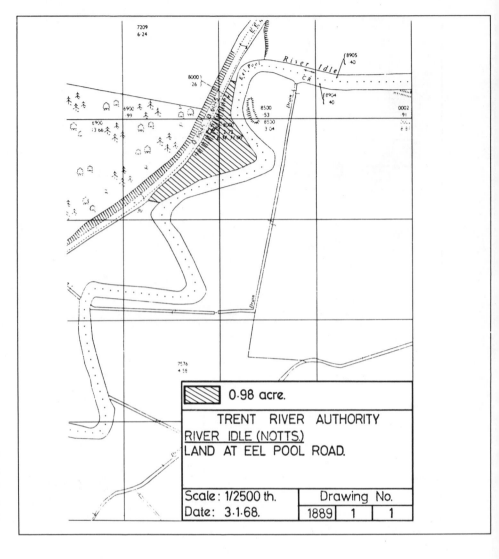

7.2   *Drawing showing the land to be pur-*
chased for a small scheme; a land plan.
It is based on a photographic reproduc-
tion of a 1:2500 Ordnance Survey map.
(Supplied by the Engineer to the Trent
River Authority)

Ordnance Survey sheets as background for this job, but the increased thickness of line, with correspondingly fuzzier edges, offset any increased accuracy obtained from the use of larger drawings.

Desk studies must be carried out carefully and to the limits of the accuracy imposed by the data used. If two sites are of relatively equal merit, rough work in the preliminary stages can lead to a wrong decision which might only be discovered when much expensive site investigation had already been carried out.

7.5 SITE SURVEYS AND INVESTIGATION
After a desk study has shown that a site is suitable, it must be investigated more fully before engineering design can begin. Few sites are properly investigated. Sometimes there is insufficient time, but often the engineer cannot persuade his employers that

they should spend their money on anything apparently so unproductive.

Subsoil investigations are slow and expensive, usually consisting of a number of carefully referenced boreholes from which samples are taken and in which *in situ* tests are made. There must, however, be uncertainty about the subsoil in the gaps between boreholes. It is incredible how many small but nasty geological faults are missed by site investigation only to be revealed once excavation for the foundations begins.

Shallow sites can be investigated by trenches or pits which enable larger samples to be taken for testing and the substrata to be seen. These are more expensive than borings and can be a nuisance when work actually commences on the site.

More faith is being placed in the results of seismic and resistivity tests for quicker, nondestructive investigations. As they become

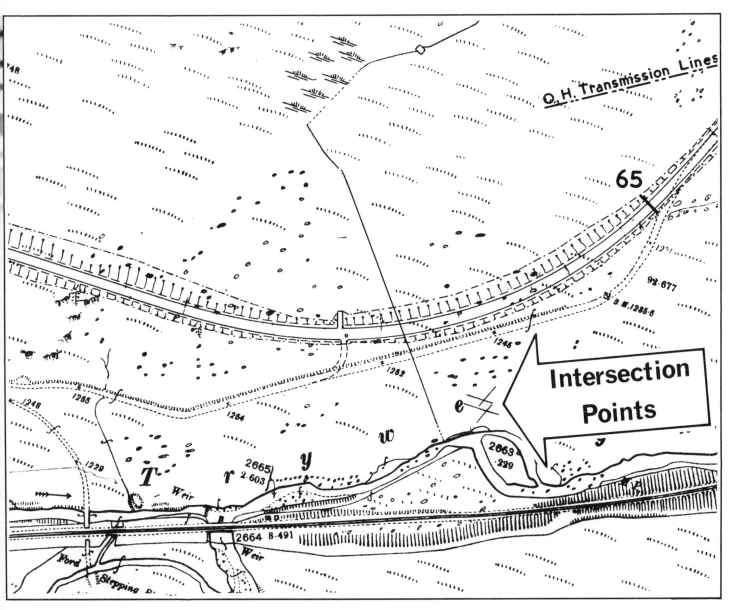

more common, greater skill is being gained in the interpretation of the results.

Most of the information gained from site investigations can be interpreted more quickly if it is presented graphically.

Typical borehole logs and reference plans, are shown in Figs. 7.4 and 7.5.

7.6 EVIDENCE TO OWN COMMITTEE

Few engineeers these days are employed by a completely autocratic individual from whom a quick decision can be obtained. More usually, the team which the engineer heads has to persuade another team (the committee), headed by a chairman, that a scheme is suitable. So many compromises are effected by both sides because of outside influences (planning and amenity considerations, land-

scaping, politics, etc.) that, unless a very careful diary is kept of this gestation period, it is difficult afterwards to pin down responsibility for a particular decision, especially if there is any recrimination. The introductory passages of many technical papers describing projects illustrate this process of debate, which is all part of the design process. Sometimes it becomes obvious that the chronicler either cannot remember the intricate steps followed at this stage or cannot reconcile them with his own ideas.

Many engineers to public authorities become skilled salesmen of ideas and know how to present their schemes attractively to their employers. In addition to facts, figures, and balance sheets, the engineer uses many graphic forms of presentation: maps, eleva-

7.3 *Alignment of a road in a mountainous area of Wales. The drawing is based on a 1:2500 Ordnance Survey map.*
(Supplied by the County Surveyor to Merioneth County Council)

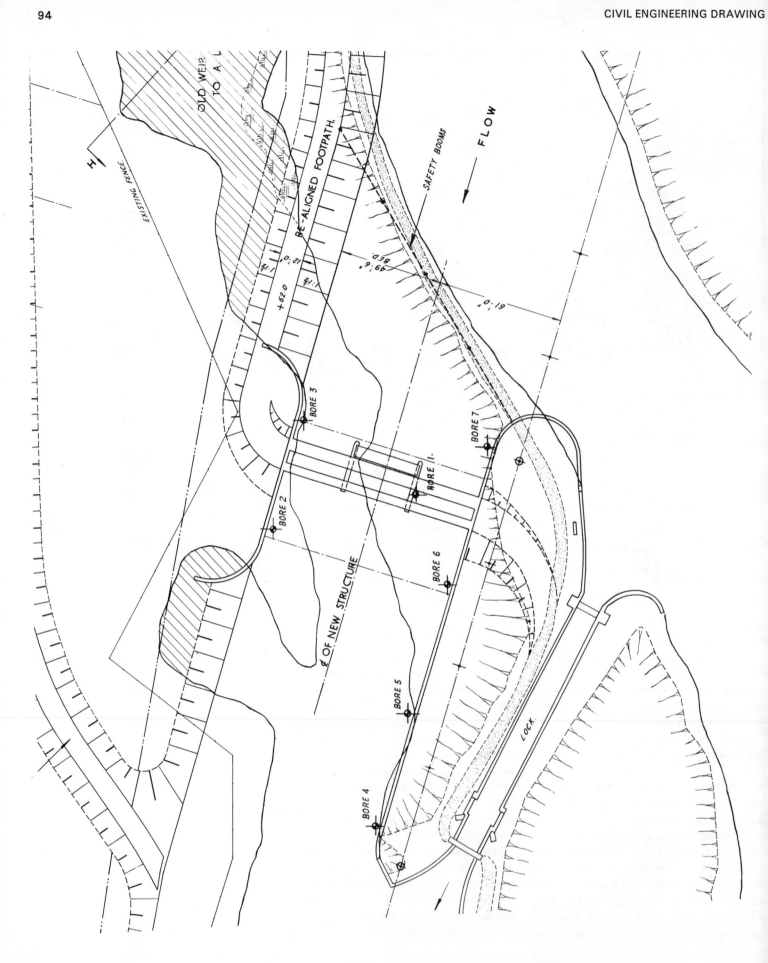

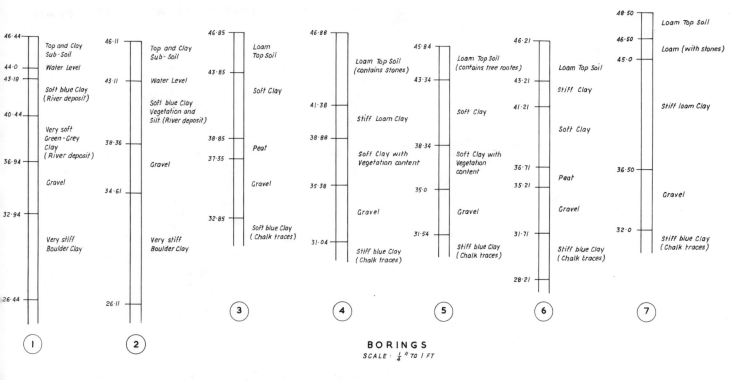

BORINGS

SCALE : $\frac{1}{4}$ " TO 1 FT

tions and sections, perspectives, and artist's impressions.

A committee's attention can be directed, by means of cartoons, to a point that the engineer wants resolved. Cartoons are drawings shorn of unnecessary detail and complication, see Fig. 7.6. They frequently accentuate a single aspect of a scheme. They should be simple, bold, and aided by the skilful use of colour.

All drawings to be used by a committee should be easy to handle. It is better to have special A4 drawings made, which can be bound into a report, rather than waste time and tempers while the members of the committee struggle to unfold and flatten larger drawings.

7.7 PARLIAMENTARY EVIDENCE

Our freedom of action is always limited, whether we are children depending on our parents' patience or a public body bounded by Parliamentary statute.

7.4 (a) Part of a working drawing showing the location of site investigation boreholes on the alignment of new sluice works. (Supplied by the Engineer to the Great Ouse River Authority. Eaton Socon Sluice Improvement)

If a civil engineering project extends to works or areas beyond the statutory limits of an authority, a special Act of Parliament or a Parliamentary Order must be sought.

An authority's submission to Parliament is called a Bill, which must be approved by the most democratic assembly possible, a town or city meeting for example, before being deposited at Westminster.

A very large number of special Acts of Parliament were sponsored by the newly formed railway companies about the middle of the nineteenth century. Each company had to prepare and give evidence before its own Parliamentary Committee with the engineer as the most important witness and coordinator. Communications then were primitive and most engineers connected with railways and similar large projects set up offices in Westminster to be close to Parliament. They could thus be called upon to organize and give evidence without wasting time at the expense of their normal professional affairs. The newly built' thoroughfare, Victoria Street, offered modern accommodation for these pioneers. It is only now, a little over a century later, that many of the successors of these early engineers are moving to the provinces, partly on account of high office rents and difficult travelling conditions. Most of the original Victorian dwellings, too, are being demolished for more intensive development.

7.4 (b) Details of the boreholes shown in (a) taken from another part of the same drawing. (Supplied by the Engineer to the Great Ouse River Authority. Eaton Socon Sluice Improvement)

7.5 Two pages from a site investigation report.
(a) Location of boreholes and site plan.
(b) Log of one of the boreholes.
(Supplied by Sarel of Watford, with permission of the Department of Highways and Transportation of the Greater London Council)

(pages 96 and 97)

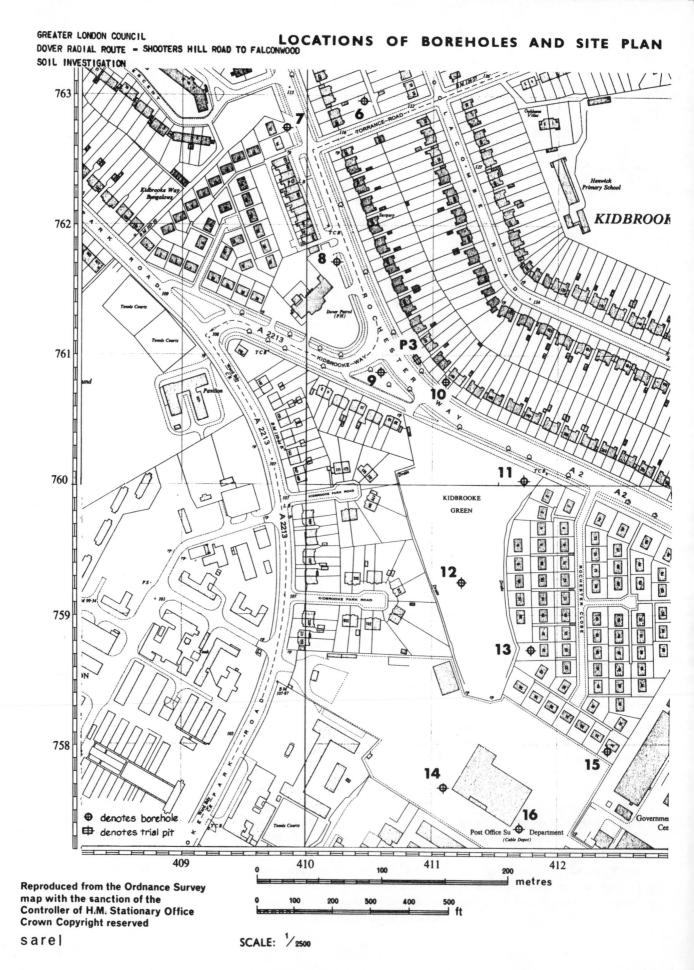

LOCATIONS OF BOREHOLES AND SITE PLAN

⊕ denotes borehole
⊞ denotes trial pit

sarel

SCALE: $\frac{1}{2500}$

GREATER LONDON COUNCIL
DOVER RADIAL ROUTE - SHOOTERS HILL ROAD TO FALCONWOOD
SOIL INVESTIGATION

RECORD OF BOREHOLE 9

Type of Boring shell and auger
Dia of boring 8in to 18.30m / 6in to 20.50m
Lining tubes 8in to 14.80m / 6in to 20.10m

Ground level 36.15m OD
Date started 6-11-68
Date completed 11-11-68

Scale ft m	Date and Depth of Boring	Depth of Casing (m)	Water Level (m)	Samples Depth (m)	Samples Type	Change of Strata Legend	Change of Strata Depth (m)	Reduced Level	Description of Strata	
	6-11-68	nil		0.30	D				MADE GROUND - mainly topsoil, ashes, firm brown and grey silty clay, sand and occasional gravel	Sand/cement
		nil 0.90	seepage	0.90 0.90-1.35	D U4(26)		1.05	35.10	Stiff brown and grey silty CLAY with pockets of brown sand and occasional fine gravel	
	1.20	1.50		1.70-2.15	U4(20)		2.15	34.00		
		1.50		2.45-2.90	U4(27)				Stiff brown and grey silty CLAY with partings and pockets of brown fine sand	
	3.50 7-11-68	1.50	nil 2.60	3.50	D & W		3.95	32.20		
		3.50		3.95-4.40	U4(32)				Very stiff dark grey silty CLAY with partings of brown fine sand	
		3.50		4.90	D					
		3.50		5.50	D		5.50	30.65		
		3.50		5.65-6.10	U4(34)				Stiff fissured dark brown-grey silty CLAY	
		3.50		7.15	D					
		3.50		7.45-7.90	U4(36)					seal
	8.40	3.50	▼	8.25	D		8.25	27.90		sand
	9.15 8-11-68	8.55 9.15	4.40 5.65	8.70-9.00 9.15	C(48) & B W				Very dense black rounded GRAVEL with some grey fine sand	seal
		9.30		9.45-9.75	C(59) & B					
		10.20		10.35-10.65	C(56) & B		10.80	25.35		
		10.95		11.15-11.35	S(100) & B				Very dense dark grey fine SAND with some shells	Excavated material
		11.75		11.90-12.20	S(129) & B					
		12.50		12.65-12.75	S(50) & B					
		13.25		13.40-13.55	S(81) & B		13.85	22.30		
		14.00 14.80		14.15-14.50 14.80	S(85) & B D		14.65	21.50	Very dense dark grey silty fine SAND with traces of shells	
	15.25 9-11-68	14.80	6.10 5.65						Cemented SHELLS	
		14.80		16.30	D		16.15	20.00		
		14.80		16.45-16.90	U4(78)				Stiff dark grey silty CLAY with shells	
		14.80		17.35	D					
		14.80		18.60	D		18.30 18.45	17.85 17.70	Cemented SHELLS	
		14.80		18.75-19.20	U4(86)				Very stiff dark grey silty CLAY with shells	
		14.80	5.80 5.35	19.65	D					
	20.10 11-11-68 20.50	20.10 20.10 20.10	5.35	20.25-20.50	U4(130)*					

END OF BOREHOLE
Depth: 20.50m
Elevation: 15.65m OD

Key
▼ denotes ground water first encountered
U4( ) ,, 4 in dia undisturbed sample
U1½( ) ,, 1½ in ,, ,, ,,
D ,, disturbed sample
B ,, bulk sample
W ,, water sample
S( ) ,, standard penetration test
C( ) ,, cone penetration test
(24) ,, number of blows
* ,, sample not recovered
▮ ,, piezometer tip

Remarks

A rock chisel was used between depths of 14.65m and 16.15m

Scale: 1 : 100 **FIG 23**

Form No. G12 (9/68)

soil and rock engineering limited

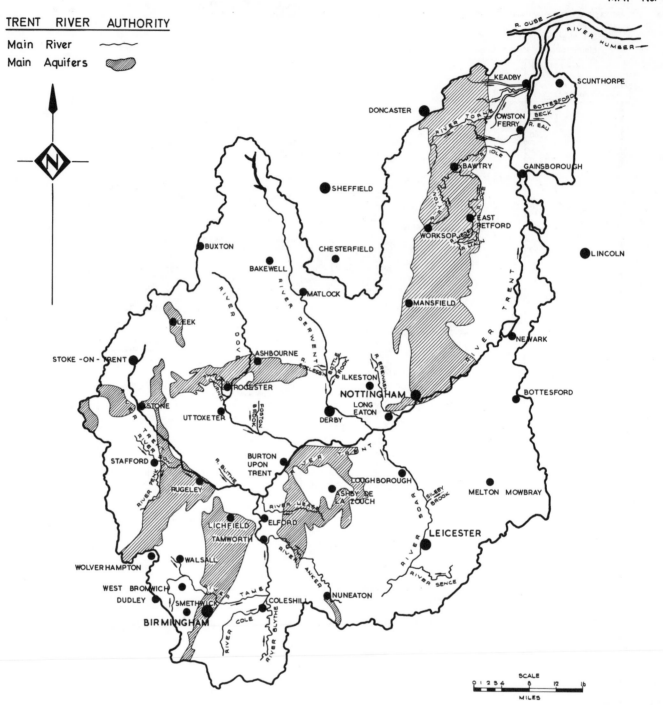

7.6   *Cartoon map of the River Trent basin from which all non-hydrological information has been omitted. (Supplied by the Engineer to the Trent River Authority)*

In addition, better communications, brought about by the railways, the aeroplane, and the telephone, themselves created by engineers, make it possible for this unique concentration of civil engineers to disperse to more amenable working areas.

As at a public enquiry so in a Parliamentary Committee, the engineer and his employer will be made very much aware of public feeling towards their project. Any move to close a railway or road brings forth voluminous and vociferous evidence that the service is operating almost to capacity and that the need is for extra facilities not closure. A remote, infertile valley, when proposed as a reservoir site, becomes a verdant paradise and home of hitherto unrealized cultures.

A Hearing before a Select Parliamentary Committee is similar, in some respects, to a court of law inasmuch as counsel wear wigs and robes and witnesses are under oath.

The case for the promoters is usually presented by learned counsel. It is he who leads the engineer and his expert witnesses through the proofs of evidence in support of the case.

Engineering evidence is supported by many drawings: maps, plans, sections, graphs, etc., copies of which are given to members of the Committee and to the opposition, who scrutinize them in great detail. Any errors or omissions are soon magnified by the opposition counsel.

The promoters of a Bill must be very flexible and capable of turning out clear, authoritative evidence, including drawings, very quickly, overnight if necessary, to meet the sudden turns and developments of the legal argument. Rapid, skilful artists and draughtsmen therefore are essential. (The term Parliamentary draughtsman, incidentally, refers to a specialist lawyer who draws up the phrases of Parliamentary documents.)

A large cartoon, showing the features of the scheme, can be of considerable assistance if hung up in the Committee Room. It helps the sometimes inarticulate engineer to present evidence to the Committee and perhaps strengthens his position against the articulate opposition counsel who may not understand drawings too well.

If the authority is lucky (or rich) enough to retain the most suitable counsel to present its case, the engineer will have more periods of elation than despair during the Hearing. After the Hearing, it is often possible to pinpoint the exact phrase or piece of evidence that turned the case from a doubtful one into a victory for the proposer.

These committees discuss so many diverse aspects of the project that one begins to wonder what connection they all have with engineering. It is salutary to realize what a vast impact engineering has on society and its environment.

The whole, sometimes protracted, business can be frustrating to the engineer who is anxious to get on with the construction of his brainchild. Few engineers however are vain enough to claim that no useful ideas came out of the presentation of their case. Few engineers emerge unchanged from examination by the opposing counsel.

7.8 TENDER DRAWINGS

The engineer must decide how much design need be carried out before the Parliamentary stage. Detailed design would be a waste of valuable time as the scheme might be changed out of all recognition. At the same time, enough general or desk design must be carried out so that the engineering difficulties can be evaluated. Witnesses must also have answers ready for all possible questions from the opposition.

Once the legal dust has begun to settle after the Act has been passed and his employers have gone off to find the money to pay for the scheme, the engineer must prepare his first formal set of engineering drawings, the tender drawings. These are the drawings which, together with the other tender documents, bills of quantities, specifications, etc., will describe the scheme to the contractor so that he can price the construction.

Curiously, large schemes often go to tender with fewer drawings and simpler documents than smaller straightforward ones. Two schemes have been known to go to tender with about sixty drawings each, one worth a few tens of thousands of pounds, the other over a hundred million pounds.

The design of small, run of the mill schemes can often be completed in one go. The engineer may have done something similar before and, therefore, stands a reasonable chance of knowing most of the answers in advance. Some slight alterations are to be expected, but may not involve more than half a dozen drawings. With this small type of scheme, the tender drawings, the contract drawings, and perhaps even the record drawings may be the same.

With very large schemes, on the other hand, many of the answers will not be known until the work has started, e.g., until some excavation has been made on a tricky site. Sometimes the sheer volume of drawing office work would occupy too much time and unnecessarily delay tendering. In any case, contractors do not want to be burdened with too much paper and detail when tendering. On the other hand, they must be acquainted with the general principles of the construction, the quantities involved, and the difficulties of the site.

The tender drawings are the first evidence to the world at large of the scale, type, and quality of the scheme involved and, as such, are prepared by many engineers and authori-

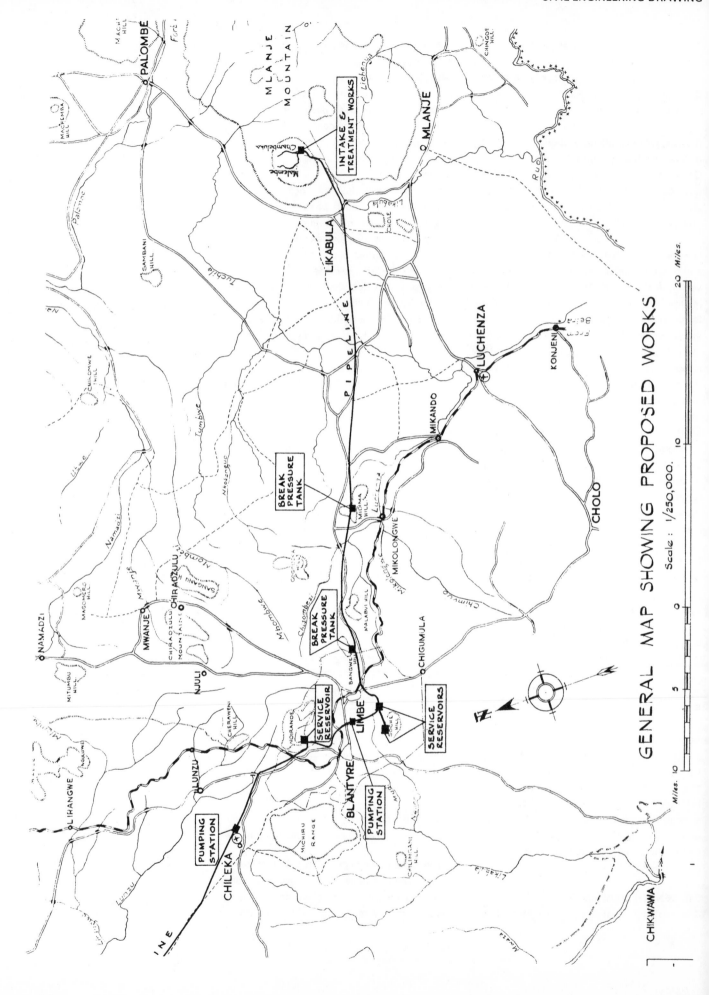

GENERAL MAP SHOWING PROPOSED WORKS

Scale : 1/250,000.

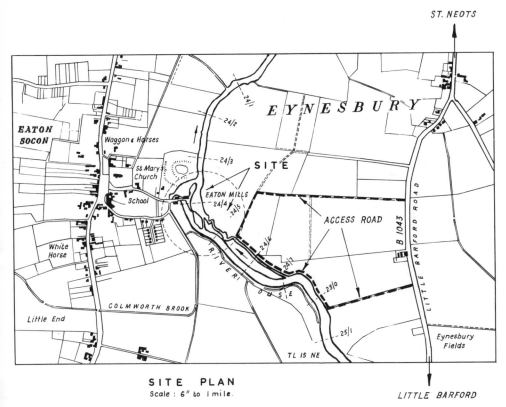

SITE PLAN
Scale : 6" to 1 mile.

7.8 *Site plan of proposed improvement works, showing site access. (Supplied by the Engineer to the Great Ouse River Authority. Eaton Socon Sluice Improvement)*

ties with considerable care, often with particular attention paid to draughting and presentation.

The contractor will hardly be concerned with the prestige value of the drawings to the promoting authority, but he will be concerned with the clarity and logical expression of the engineering detail and its accuracy.

There must be no ambiguity. Probably the only way of ensuring this is to ask an engineer, who has had no previous experience of the scheme, to examine the drawings critically and to say whether he can obtain a clear impression of the scheme and its detail. Junior engineers sometimes feel that their boss is overcritical when he finds faults in something they have worked hard on for weeks. But it is his experienced eye, coming, without preconceived ideas, to the drawing, that has spotted the weaknesses. It is frightening how mistakes or anomalies can escape detection. It can be a salutary experience for any design engineer to have another engineer read and interpret a bar bending

7.7 *1:250 000 map showing the essentials of a recommended scheme for the water supply to two African towns. (Supplied by Scott & Wilson, Kirkpatrick & Partners)*

schedule that he has prepared for a reinforced concrete design, see section 8.3, and to hear his comments afterwards.

Tender drawings can encompass every scale, convention, and concept used in engineering. If the project is hidden away in an obscure corner of the world, the pictorial story can start with a sheet of maps showing all or part of a continent, going on to, *inter alia*, location maps or plans (see Figs. 7.7, 7.9, and 7.10), access plans (see Fig. 7.8), general arrangement drawings (see Fig. 7.11), and finally, drawings with constructional detail (see Figs. 7.12, 7.13, 7.14, and 7.15). Intimate constructional details are included only for smaller or routine schemes. However, all knowledge of the site: borehole logs, geological sections, cross sections and longitudinal sections, closely contoured plans, etc., will be needed, however large or small the job.

7.9 CONTRACT DRAWINGS

For several weeks, or months, (if the contractors are fortunate) after the rush of completing the tender drawings and compiling the documents for issue to competing contractors, the engineer will have some respite during which he can carry on with detail design.

During the tendering period, contractors

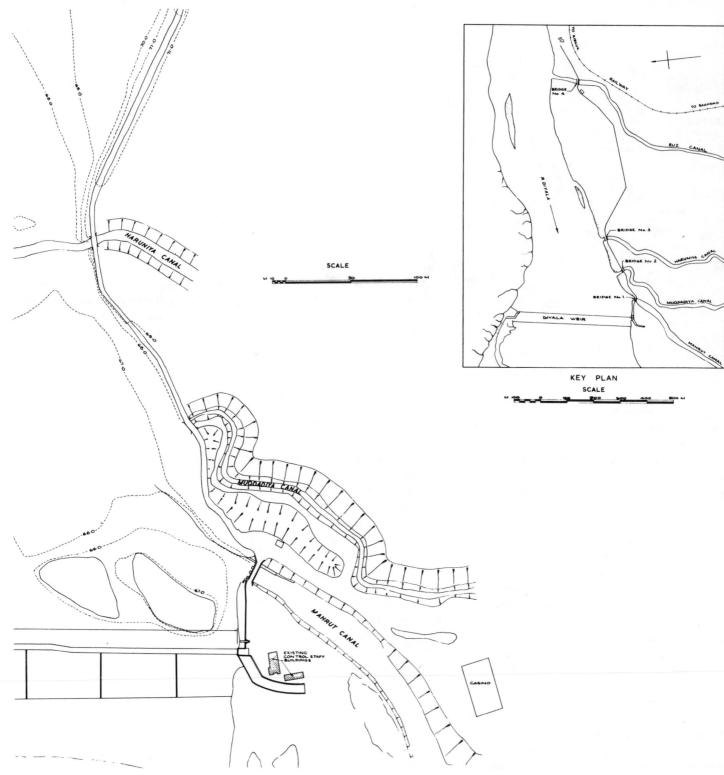

7.9 Key or location plan of irrigation works
in the Middle East, together with part of
the site survey to a larger scale. (Supplied by Sir M. MacDonald & Partners)

7.10 A continuation of Fig. 7.9. Plan and
longitudinal section of the headworks of
the irrigation canal, with all relevant data
tabulated concisely, accurately, and
neatly below the section. (Supplied by
Sir M. MacDonald & Partners)

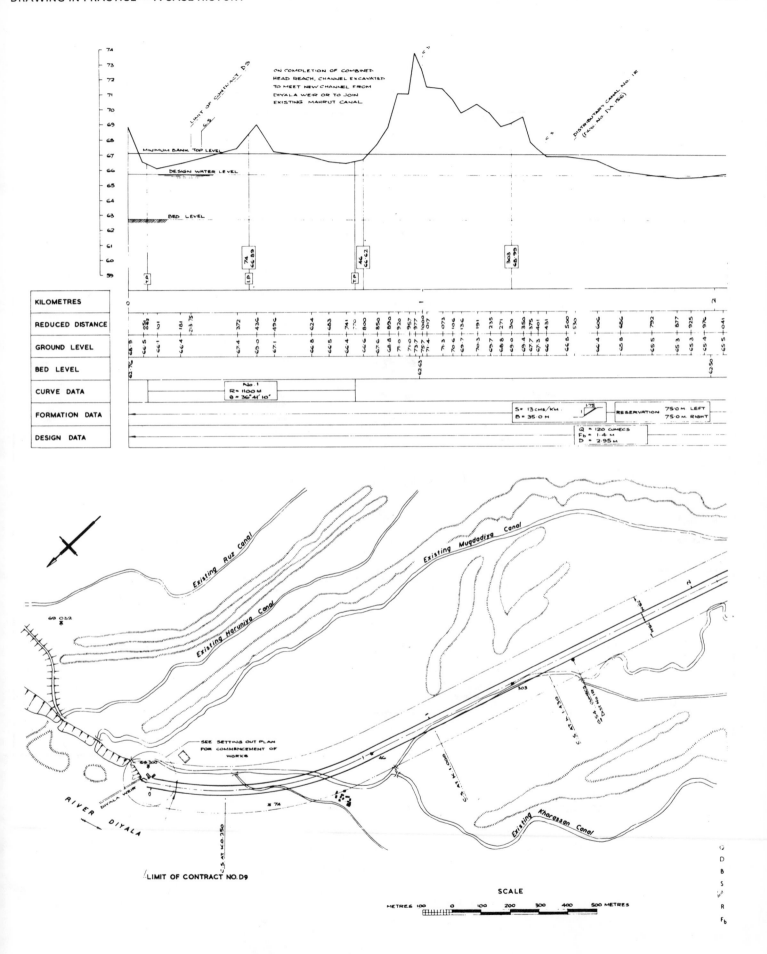

ON COMPLETION OF COMBINED
HEAD REACH, CHANNEL EXCAVATED
TO MEET NEW CHANNEL FROM
DIYALA WEIR OR TO JOIN
EXISTING MAHRUT CANAL

MINIMUM BANK TOP LEVEL

DESIGN WATER LEVEL

BED LEVEL

KILOMETRES									
REDUCED DISTANCE									
GROUND LEVEL									
BED LEVEL									
CURVE DATA									
FORMATION DATA									
DESIGN DATA									

No. 1
R = 1100 M
θ = 36° 41' 10"

S = 13 CMs/KM   1:75
B = 35.0 M

RESERVATION   75.0 M LEFT
              75.0 M RIGHT

Q = 120 CUMECS
Fb = 1.4 M
D = 2.95 M

Existing Ruz Canal

Existing Muqdadiya Canal

Existing Haruniya Canal

Existing Khorassan Canal

SEE SETTING OUT PLAN
FOR COMMENCEMENT OF
WORKS

RIVER DIYALA

DIYALA WEIR

LIMIT OF CONTRACT NO. D9

SCALE

METRES 100   0   100   200   300   400   500 METRES

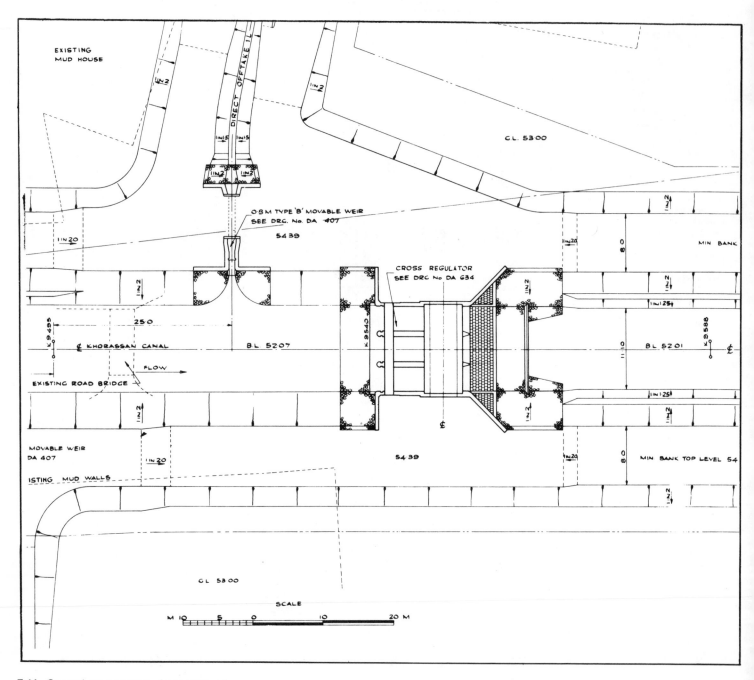

7.11 *General arrangement of a regulator (con-*
trol sluice) in an irrigation canal.
(Supplied by Sir M. MacDonald &
Partners)

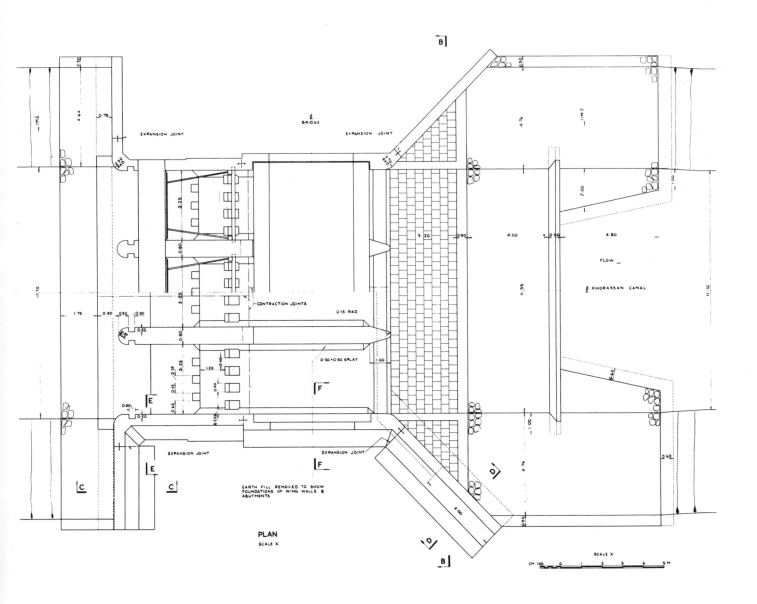

PLAN
SCALE 'A'

7.12 General arrangement of the canal regulator in Fig.7.11 to a larger scale showing constructional details and dimensions, see also Fig. 7.13. (Supplied by Sir M. MacDonald & Partners)

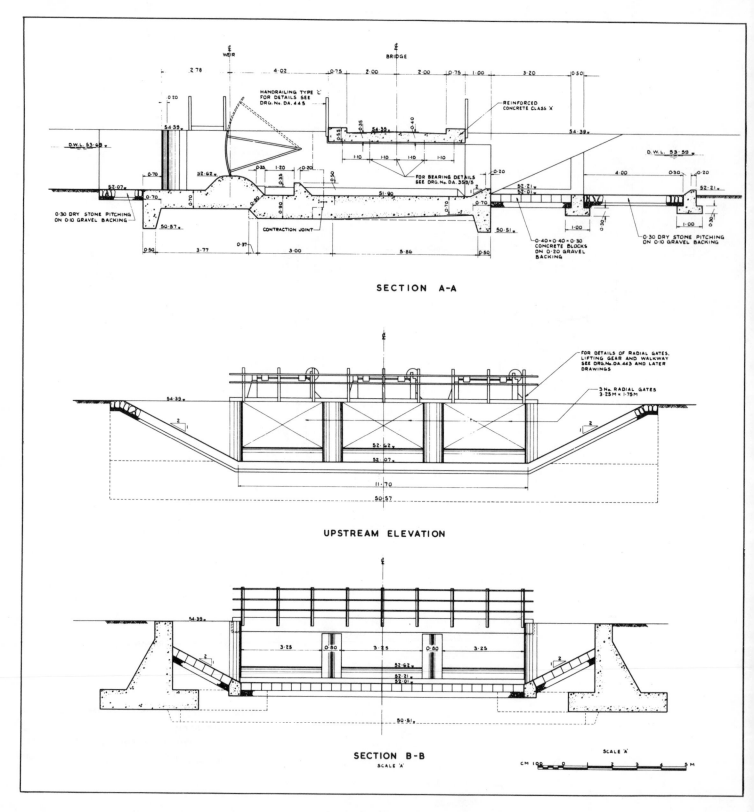

SECTION A-A

UPSTREAM ELEVATION

SECTION B-B
SCALE 'A'

SCALE 'A'

7.13 Longitudinal section, upstream and downstream elevations of the canal regulator shown in Fig. 7.12. See how the section reveals the construction and layout almost at a glance. (Supplied by Sir M. MacDonald & Partners)

7.14 Plan, longitudinal and cross sections of an accommodation (access) bridge over an irrigation canal. The longitudinal section reveals the construction almost at a glance. Notice that Section B-B is distorted. See its alignment on the plan,

(Supplied by Sir M. MacDonald & Partners)

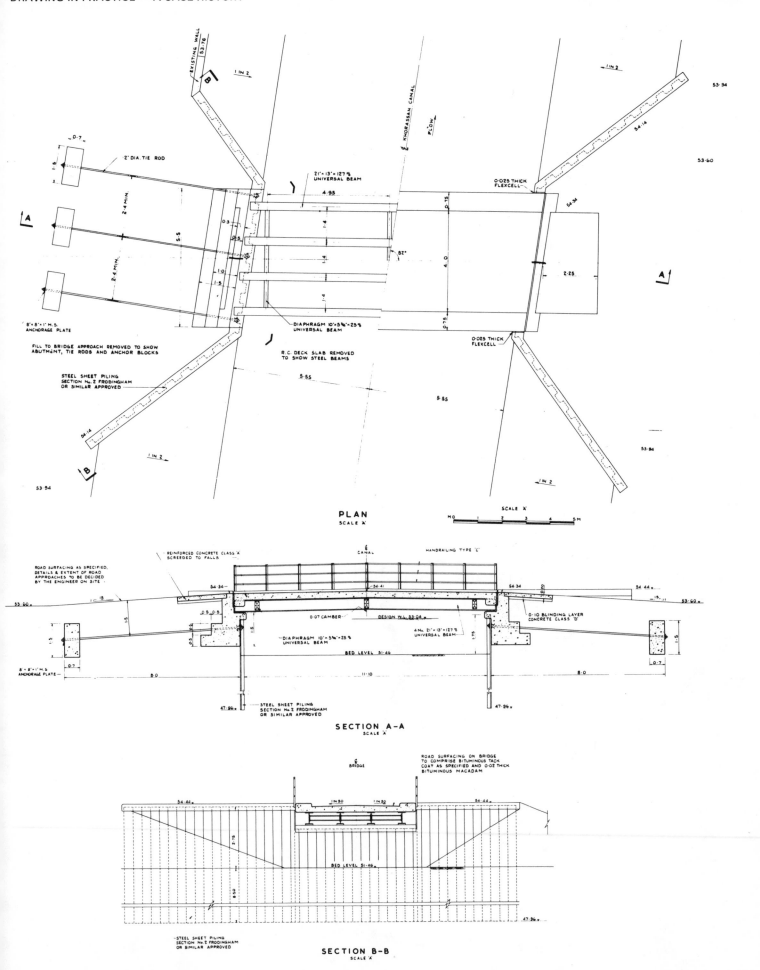

PLAN
SCALE 'A'

SCALE 'A'

SECTION A-A
SCALE 'A'

SECTION B-B
SCALE 'A'

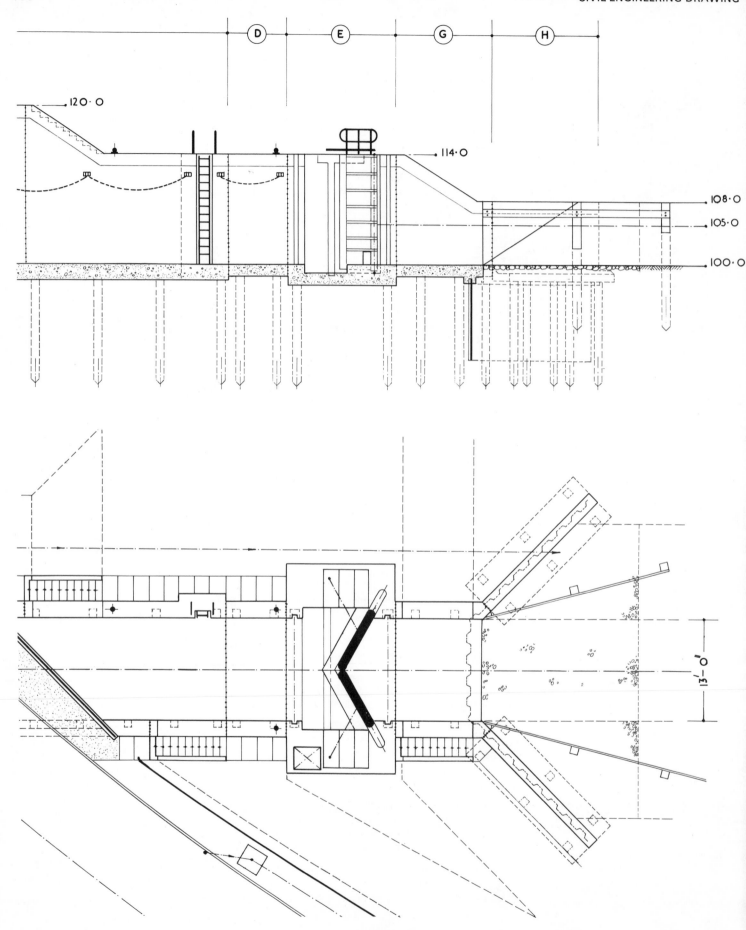

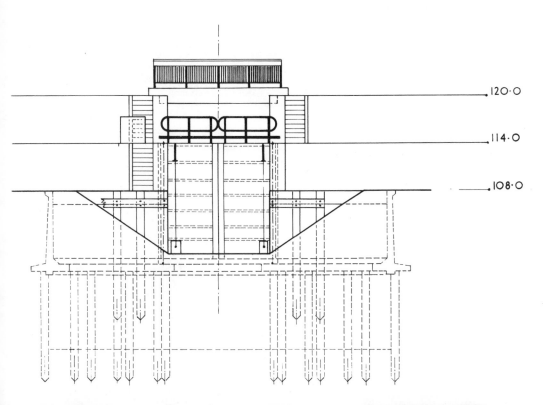

120·0

114·0

108·0

7.15 Detail of a navigation lock.
(a) Part of the longitudinal section and plan from the general arrangement of the lock (Supplied by the Engineer to the Great Ouse River Authority. Hermitage Lock Improvement).
(b) View from below the lock.
(c) View from the road bridge.

will be interpreting the tender drawings and preparing their bids. It sometimes happens that one contractor may be skilled in a particular method of construction that is different from that shown in the tender drawings. He may submit a tender based upon his speciality even though it will necessitate an alteration in the design, or even the concept, of the scheme. If his price is competitive and the engineer agrees with his proposals, modified drawings will have to be prepared. The engineer must be flexible in his ideas and realize that there may be better solutions to the problem than his own.

Once the contractors have submitted their tenders, the engineer will have to evaluate and compare them, to look for anomalies in pricing, to see whether the proposed constructional methods are suitable, to see if any legal loopholes have been prized open, etc. On the basis of the engineer's report and recommendation, his employer will accept one of the tenders, for which a legally binding contract will then be prepared, of which the contract drawings form a vital part.

If tendering has been straightforward, the contract drawings will be the same as the tender drawings. If alternative proposals have been accepted, new or additional drawings will have to be prepared.

Contract drawings are usually printed on a linen based paper so that they will survive long storage in the employer's and contractor's archives.

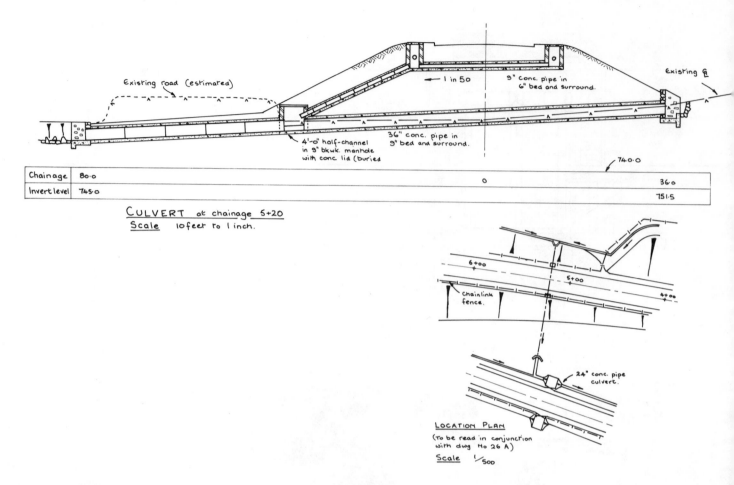

Chainage	80·0		0		36·0
Invert level	745·0				751·5

CULVERT at chainage 5+20
Scale   10 feet to 1 inch.

LOCATION PLAN
(To be read in conjunction
with dwg No 26 A)
Scale   1/500

7.16 A working drawing, in pencil on tracing paper, of drainage details of a new road. The drawing was made quickly in a small site office to meet the contractors's construction programme. (Supplied by the Water Engineer, City of Liverpool Water Department)

7.10 WORKING DRAWINGS

The working drawings fill the gaps in the constructional detail left by the general nature of the tender drawings. As the name implies, they are for working or building purposes and must, therefore, represent the engineer's last thoughts on construction. Drastic changes of mind hereafter can be expensive, especially if the contractor is already building and is breathing down the engineer's neck for more detail.

The engineer can sometimes hand over the small routine job to the Resident Engineer on site since the tender drawings contain all the constructional detail. If amendments are small, the site staff can often cope with them, although small contracts may not have sufficient site staff to manage major alterations to drawings.

For the large project, which had only outline drawings at the tender stage, the busiest time for the drawing office lies ahead. The drawing office may be at the engineer's headquarters or it may equally well be on site.

Site or working drawings still need to be complete, accurate, logical, and unambiguous, but the need for prestige completion and presentation has disappeared. The work will probably be left in pencil on tracing paper or polyester film, with notes and instructions written in very freely, but legibly, see Fig. 7.16.

This is perhaps one of the most testing and rewarding aspects of engineering drawing. The engineer knows that the drawing is required urgently, that he must get it right first time, while his working conditions may be anything but ideal.

7.11 COMPLETION, RECORD, OR 'AS-BUILT' DRAWINGS

By the very nature of civil engineering, where the engineer is delving into the unknown or partially known recesses of the earth and sometimes designing to the limits of technology, it is unlikely that everything will go literally according to the plan. There is bound to be some alteration or addition to even the very latest working drawing due to some vagary of the site or of the client.

However trifling the deviation from the working drawing, it must be recorded. Every alteration must be recorded on a set of completion or record drawings.

All this work can be accomplished at one sitting for a small job where all has gone well and the engineer in charge is not likely to be whisked away to another job. But for a large project, where each phase or section can take months to complete and be worth millions of pounds, the record drawings must be prepared simultaneously as the work proceeds.

The simplest version of a completion drawing will be a set of prints of the tender or workings drawings duly amended in red ink or pencil. The more involved alteration could well lead to a completely new drawing or set of drawings.

The preparation of completion drawings can be a test of the site engineer's personal concentration and devotion. He may have been left in charge of winding up the job and doing the completion drawings at the same time, when the only real incentive may be to get off the lonely site and back to work on something creative. He should try to realize the value of his work and drawings to the maintenance engineer who, later on, may need to find some buried detail.

7.12 ILLUSTRATIONS IN TECHNICAL PAPERS

The final phase in the graphical representation of an engineering project often comes in technical papers. These are published in journals which serve to pass on the useful knowledge gained or new techniques used. This phase may come ten years after the work was first started in the design office.

Unless the paper is to be illustrated by means of full size drawings, which will necessitate the use of unwieldy, expensive, and infuriating pull-out sheets, the drawings will be photographically reduced. Although some designers find albums of the drawings of their previous projects, photographically reduced to about half size (a quarter of the area), are very useful for general reference, much of the detail is lost and notes in the smallest printing are illegible. In any case, too much construction detail is shown on ordinary drawings to be of interest to most readers of technical papers. Even graphs and trend charts need special preparation, cartooning almost, to achieve maximum clarity.

The drawings used to illustrate papers in the *Proceedings of the Institution of Civil Engineers* are a good example of simplified but straightforward technical illustration, see Fig. 7.17. As the small page size of the *Proceedings* restricts illustrations to a width of about 100 mm, a reduction of five or six times is not uncommon. Hence, although authors submit specially made, simplified drawings, the Institution's own draughtsmen usually have to thicken some lines to ensure clarity. Moreover, very thin lines are too frail on the printer's plate which must be capable of printing nearly forty thousand copies. (An offset printing process maintains the clarity of very thin lines better than plates.) Captions are prepared by the Institution to suit the degree of reduction. They are set in type, printed, cut out, and stuck in place on the original drawing, as shown in Fig. 7.17. The smallest acceptable size for lowercase letters on the finished page is about 1 mm high (6 point). For example, for a reduction by five, the captions for the original are prepared in 30 point letters. A whole page of 6 point print would be unacceptable, but isolated notes surrounded by blank paper are clear at this size.

Some journals have developed a stylized technique of their own. For example, the drawings used in *Water Power* are peculiarly and attractively their own, with outlines made up of thick and thin lines which seem to suggest depth and substance in a subtle way.

7.13 SLIDES

Slides used to illustrate lectures must be made from specially prepared drawings if they are to have any impact. The overwhelming detail of the typical engineering drawing is confusing, if not completely invisible, on the screen. If in doubt about its suitability as a slide, view the original drawing from a distance comparable with that of an audience at the back of a hall 10 m long viewing a screen 1.5 m wide, and see if all lines and captions are clear, i.e., a drawing 400 mm wide should be viewed from a distance of

$$10 \times \frac{0.4}{1.5} = 2.67 \text{ m,}$$

and an A1 drawing from about 5 metres. The thinnest line on a drawing 400 mm wide should be about 0.5 mm thick which, as we saw in section 3.5 which refers to BS 308, gives rise to thick lines a millimetre or more wide—much thicker than the inexperienced

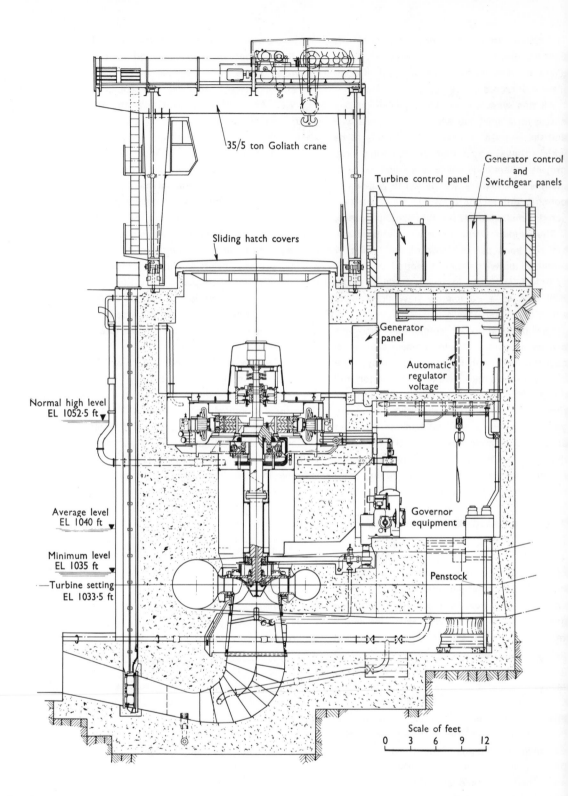

35/5 ton Goliath crane

Generator control and Switchgear panels

Turbine control panel

Sliding hatch covers

Generator panel

Automatic regulator voltage

Normal high level EL 1052·5 ft

Average level EL 1040 ft

Governor equipment

Minimum level EL 1035 ft

Turbine setting EL 1033·5 ft

Penstock

Scale of feet
0   3   6   9   12

7.17 *Typical section of a hydroelectric power station, specially prepared for publication in the* Proceedings of The Institution of Civil Engineers. *(Supplied by the Institution of Civil Engineers from the paper* Nkula Falls hydroelectric scheme initial development *by R. S. Arnott, D.D.A. Piesold, and J. G. Wiltshire)*

draughtsman would expect. Round, open lettering with bold lines, like the BS 308 example in Fig. 3.11 (a), is clearer than narrower, perhaps more elegant, lettering with considerable variation in line width. Script from most typewriters is unsuitable as the type is thin compared to the size of the letters.

Coloured lines on drawings reproduced on slides present a special problem, since the flood of light on to the screen representing the white background degrades the colours and will almost kill pale colours. Only skilled photographers, with carefully selected colour materials and lighting, can reproduce coloured line work really satisfactorily; the result is so often either quite healthy colours on a dingy blue-grey background or pallid lines on a clear white background.

Some workers prefer to project negative images of line work, i.e., white lines on a dark screen. This can be less tiring to the viewers than a brilliantly lit white screen, but the lecture room is then too dark for making notes without considerable background illumination. Selective colouring of lines is easily achieved by the use of felt or fibre-tipped pens on the emulsion of the slide.

Chapter 8   Some specialist applications

A modern cathedral built in three years instead of several centuries. The Metropolitan Cathedral of Christ the King, Liverpool's Roman Catholic Cathedral. (Photograph supplied by the Cement and Concrete Association)

8.1 INTRODUCTION

At the beginning of the present technological age, engineering was either civilian or military. The rapid increase in knowledge during the nineteenth century so diversified the work of the engineer that the electrical and mechanical aspects broke away to diversify even more. This still left the civil engineer with a very wide range of work. The widening of civil engineering itself has continued until the only common bonds between some of its branches today appear to be mathematics and drawing. Even the latter has developed specialist techniques and applications.

In addition to his fellow specialists, the civil engineer will often be working with mechanical, electrical, and electronic engineers, not to mention geologists, biologists, chemists, architects, etc.

In this chapter, a few specialist applications of drawing are discussed. It is intended only to introduce you to these topics so that you will be more readily able to read such drawings and appreciate more fully the overall picture of civil engineering.

8.2 ARCHITECTURE

(a) The role of the architect

Popular ideas of the role of the architect cover the extreme ends of the spectrum of his work, but usually leave blank his major role. Some architects are concerned mostly with private dwelling houses, while a few are involved with prestige structures, for example, cathedrals and tower office blocks. In the main, architects are concerned with providing the right sort of building for their clients, to which end they must correlate such aspects as planning, landscaping, ergonomics, and aesthetics.

Planning. Planning means using the site to the client's best advantage within the planning bylaws and restrictions which have been drawn up to prevent exploitation of people and localities by ruthless developers. The overcrowded, back to back housing thrown up in the industrial north during the nineteenth century, or even the straggle of houses along main roads that occurred in the nineteen-thirties, show what can happen if uncontrolled development takes place.

Landscaping. Landscaping is concerned with fitting the scheme into the existing scene as harmoniously as possible or creating an attractive environment round an alien structure.

As projects become larger and larger, the landscaper is tested more and more. Landscaping need not be expensive. It can be achieved by sympathetic attention to detail, for example, an unwanted dry stone wall, which would otherwise run downhill into a new reservoir, should be removed right back to the boundary wall rather than be left as a derelict reminder of the past use of the valley.

Ergonomics. Ergonomics is concerned with making the building fit the people who will inhabit it, in respect of both comfort and efficiency, whether the building is for domestic or industrial purposes.

Aesthetics. Aesthetics are concerned with providing a tasteful building or environment that fulfils its function efficiently. There is an obvious connection with landscaping. Nothing can hide the vast electricity generating stations that must spread across the country as we demand more power, but if care is taken in locating the 120 m high cooling towers and the 200 m high chimneys, the effect is less disastrous. The present trend, introduced for technological reasons, of using one larger multiple chimney rather than several smaller ones, creates a less fussy outline.

Architects are popularly supposed to be weak in the field of structural design while engineers reputedly have philistine views and inferior aesthetic ability. Even though generalizations, these notions are unfortunately true to some extent. The better architects and engineers do complement each other, while the best could perform each other's tasks. Two good heads are always better than one; the philosophical and technological interchange of ideas between architect and engineer, arguing from different viewpoints, will always benefit a job.

As an engineering student, you should not forget that design is aesthetic and ergonomic as well as structural. It is concerned as much with appearance and convenience as with sizes or structural members and the stresses in them.

Whether the engineer employs the architect or *vice versa* depends largely on the content of the job. If it is essentially building (in the wider sense, not dwelling houses), the architect will have overall control and will call in structural and service engineers to advise and carry out their part of the design. Bridges,

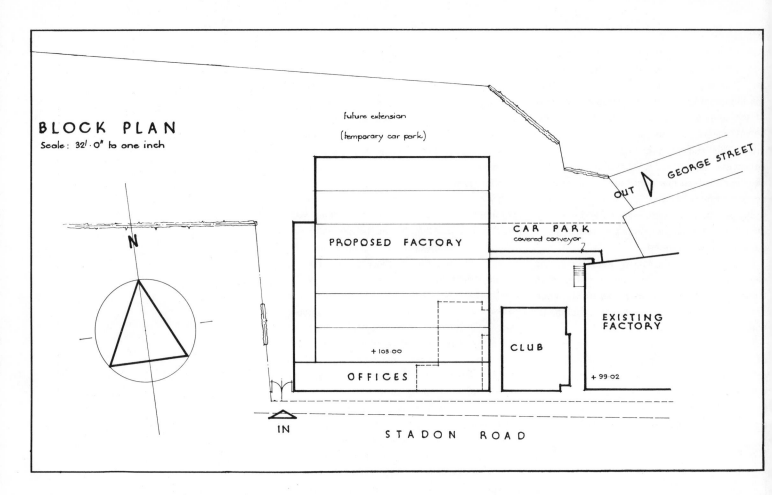

BLOCK PLAN
Scale: 32'·0" to one inch

future extension
(temporary car park)

PROPOSED FACTORY

OUT GEORGE STREET

CAR PARK
covered conveyor

EXISTING
FACTORY

CLUB

+ 103·00

+ 99·02

OFFICES

IN

STADON   ROAD

N

8.1   Block plan, from 32 feet to 1 inch scale original drawing (1:384). (Supplied by Francis W. Keyworth, LRIBA, of Melton Mowbray)

dams, power stations, motorways, etc., are engineering projects which need the advice of an architect to make them fit into their environment. Even a simple roadway cutting can benefit from landscaping.

(b) Architects' drawings

The architect builds up his designs in the same reasoned and logical way as the engineer. He starts by outlining the building on 1:100 (8 feet to 1 inch) or smaller sketch plans, which are correct scale drawings and include elevations and sections, see Fig. 8.2.

The sketch plans enable him to prepare the 1:500 scale block plans, see Fig. 8.1, required to obtain outline planning approval, i.e., to discover whether the scheme can go ahead or not.

Further work on 1:100 scale drawings leads on to 1:25 (2 feet to 1 inch) scale detail drawings which should reveal most of the snags and provide constructional working drawings, see Fig. 8.3 and Fig. 8.4.

Final planning approval on appearance, etc., is now obtained.

Construction is sometimes carried out from heavily annotated 1:100 scale drawings, but

this places considerable reliance on the building contractor and his craftsmen. This frequently causes delays on site while the architect is contacted to sort out a problem and can result in expensive remedial work on detail not revealed by the small drawings.

Perhaps the most outstanding difference between an architect's drawings and an engineer's is the realism the former obtains by the clever use of lines of different thickness and close attention to detail, i.e., double lines, however fine, to show door and window frames, shadows, etc. His buildings seem related to their surroundings and firmly attached to the ground, helped by figures, trees, vehicles, and other detail however representational.

8.2   Sketch plan from 8 feet to 1 inch scale original drawing. (Supplied by Francis W. Keyworth, LRIBA, of Melton Mowbray)

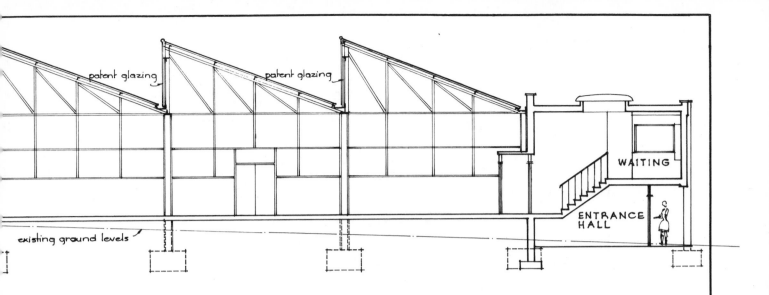

SECTION (North - South)

patent glazing

patent glazing

WAITING

ENTRANCE HALL

existing ground levels

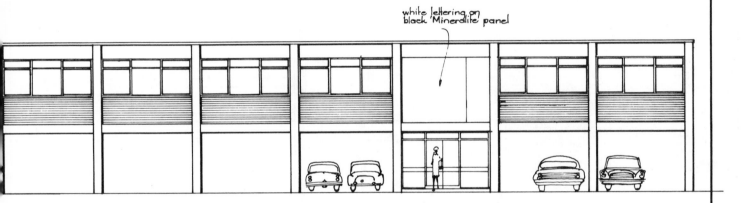

white lettering on black 'Mineralite' panel

ELEVATION TO STADON ROAD

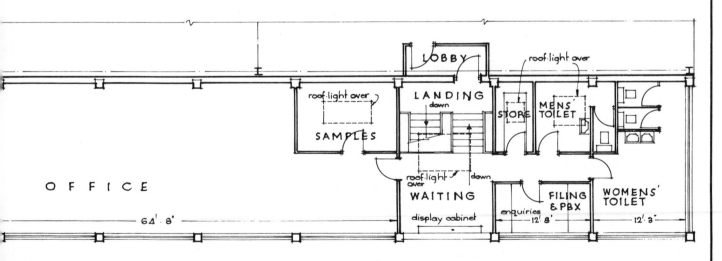

LOBBY

roof-light over

roof-light over

LANDING
down

STORE

MENS' TOILET

SAMPLES

OFFICE

roof-light over   down

WAITING

display cabinet

enquiries

FILING & PBX

WOMENS' TOILET

64' 8"

12' 8'

12' 2"

FIRST FLOOR PLAN

A

Scale: Eight feet to one inch
Date: August 1967

Francis W. Keyworth, L.R.I.B.A
Chartered Architect,

15"x 3" Art. stone weathered & twice throated
coping bedded on approved d.p.c. felt.

24"x18"x3" Art. stone
corbel.

Metal windows types

ZND5   ZND5   ZND5

ZND1   ZND1   ZND1

7"x 3" Art. stone
jamb.

7"x 4"   Art. stone
mullions

ZND5   ZND5   ZND5

155
ing.

stone

Aluminium edging strip to receive asphalte

¼" polished plate glass.

See 1" detail drawings
of this Entrance Screen

ND II FS.

Brindle brick on edge
coping with creasing tiles

¾" Macadam on 2¼" tarmacadam
on 6" hardcore.

1'9"x 6" foundation
concrete

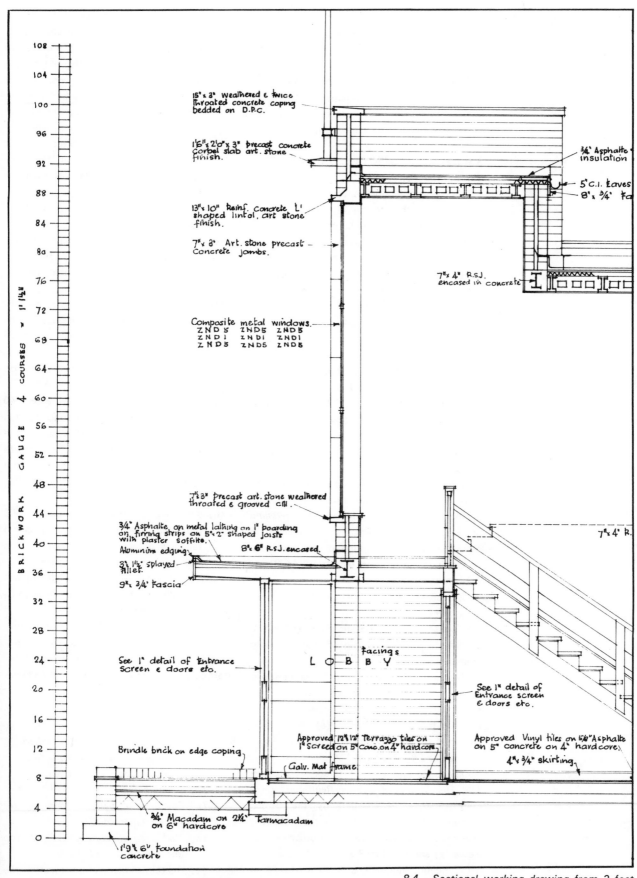

8.3 *Elevation from 2 feet to 1 inch scale original drawing; a working drawing. Prepared and used in conjunction with Fig. 8.4. (Supplied by Francis W. Keyworth, LRIBA, of Melton Mowbray)*

8.4 *Sectional working drawing from 2 feet to 1 inch original. (Supplied by Francis W. Keyworth, LRIBA, of Melton Mowbray)*

Architects do not achieve realism in their drawings any more easily than engineers do. The apparent sloppiness of some architectural drawings is misleading. There is usually an economy and clever use of line, gained only through experience and practice. Economy of line is not leaving out lines at random, but knowing which ones to leave out.

8.3 DETAILING REINFORCED CONCRETE

(a) Introduction

This section is not concerned with stress analysis and the design of reinforced concrete, but with the clear and concise presentation of its details in drawings.

Over the years, reinforcement has become more and more complicated. Individual designers and firms have developed their own methods both of shaping the bars and showing them on drawings.

BS 1478 was issued in an attempt to reduce the infinite number of possible bar shapes by suggesting that engineers should use a combination of the preferred shapes to satisfy their designs. Even in the offices that adopted BS 1478, many different drawing conventions have grown up in an attempt to describe the complicated cages of steel bars that are often necessary.

In 1968 a committee, formed by the Concrete Society and the Institution of Structural Engineers, published its report, *The Detailing of Reinforced Concrete*, in an attempt to rationalize the presentation of reinforcement in drawings. The Report adopts BS 1478.

(b) An example

Figure 8.5 shows a simple cantilevered loading wharf for lorries.

A possible loading is shown at (b) and a simplified envelope diagram of bending moments at (c).

The reinforcing bars of circular section mild steel, theoretically necessary to carry the tensile forces in the slab, are shown as heavy lines on the general section (a). A practical layout of the reinforcement is shown in section (e) and by means of half plans (d).

Experience and CP 114 (Code of Practice 114: *The Structural Use of Reinforced Concrete in Buildings*, issued by the British Standards Institution), dictate the following differences between the reinforcement in views (a) and (e), Fig. 8.5

(a) Reinforcement has been curtailed in areas of reduced bending moment.

(b) The ends of the bars are hooked to ensure that they do not slip under load.

(c) Distribution steel has been provided to maintain the design spacing of the main reinforcement before the concrete is placed. This also helps spread stresses due to concentrated loads and control stresses due to thermal expansion.

(d) Saddles have been provided to support the steel in the top of the slab (especially at the point of maximum moment over the support) until the concrete has been placed and has hardened.

(e) The upstand at the left-hand end has been reinforced with rectangular stirrups which also support the main reinforcement.

(f) Dowels have been provided to prevent the slab moving on the front supporting wall due to thermal creep or vibration. (Other methods such as keyways in the top of the wall would do as well.)

(g) No reinforcement is placed within 40 mm of the surface of the concrete, this is known as the cover. Steel at the bottom of the slab is laid on concrete blocks 40 mm thick and about 40 mm square. Top steel is carried on saddles (bar 1605).

(h) All main reinforcement is carried 300 mm past the point where it is no longer required to carry tensile loads (CP 114), the anchoring length.

Figures 8.6 and 8.7 show the reinforcement for the simple wharf slab detailed in accordance with the Report.

The necessary information is on two separate drawings. Figure 8.6 shows the arrangement of the reinforcement with as little other information as possible, while Fig. 8.7 shows

8.5 *A reinforced concrete loading wharf.*

(a) General section showing theoretical reinforcement required.

(b) Loading.

(c) Envelope of bending moments for reinforcement design.

(d) Half plans showing steel in the top and the bottom of the slab.

(e) Section showing arrangement of the reinforcement as placed.

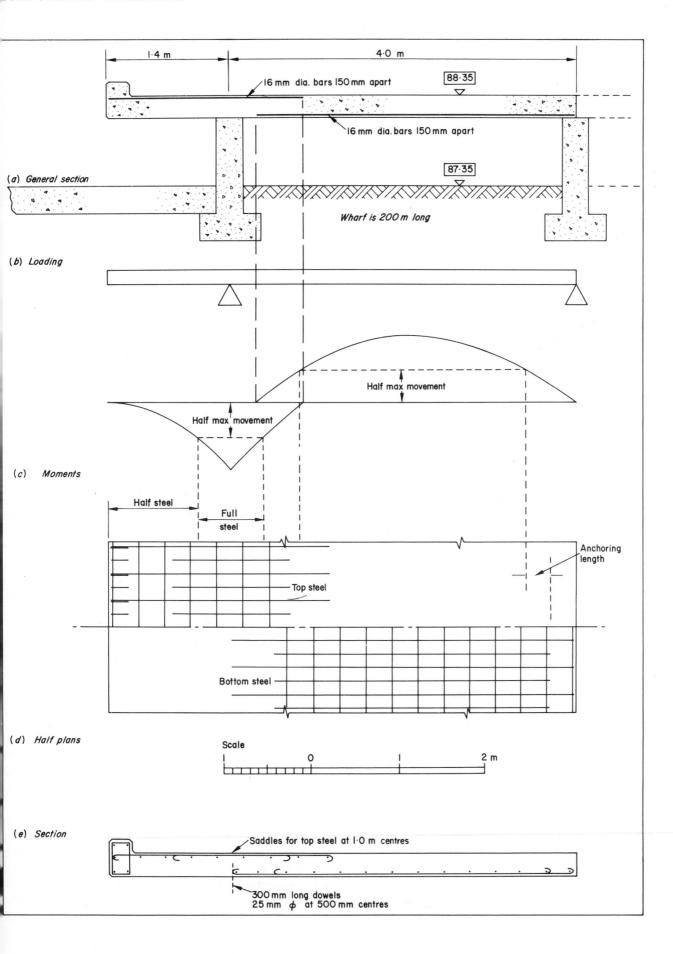

(a) General section

16 mm dia. bars 150 mm apart

16 mm dia. bars 150 mm apart

88·35

87·35

Wharf is 200 m long

(b) Loading

(c) Moments

Half max movement

Half max movement

(d) Half plans

Half steel

Full steel

Top steel

Anchoring length

Bottom steel

Scale

0   1   2 m

(e) Section

Saddles for top steel at 1·0 m centres

300 mm long dowels
25 mm φ at 500 mm centres

1·4 m

4·0 m

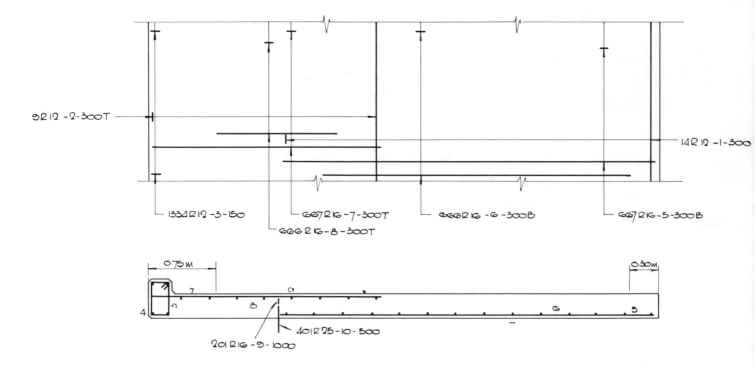

9R12 -2-300T

14R12 -1-300

1334R12 -3-150

667R16 -7-300T

666R16 -8-300T

666R16 -6-300B

667R16 -5-300B

0·75M

0·30M

401R25 -10- 500

201R16 -9- 1000

8.6 *Reinforcement drawing in accordance with the report*, The Detailing of Reinforced Concrete.

a bar schedule which enumerates and details every bar. Notice that in Fig. 8.6 no slab dimensions are given; only one bar of each type is shown fully; bars in the top and bottom of the slab are shown as full lines; and the quantity of each bar is shown once only.

The bar schedule is prepared on A4 sheets. It must be complete in itself so that a contractor can send it to a reinforcement subcontractor or to his own yard for the steel to be ordered and bent without reference to any other drawing.

Two types of bar schedule are shown in Fig. 8.7, both in accordance with BS 1478. In (a) the shape of the bars is shown pictorially (although schematically), while in (b) the shape code from BS 1478 is given instead, with standard dimensions. Schedule (b) is intended when a computer is used to prepare orders for the steel bars and for accounting. Schedule (b) can be compiled more quickly than (a) as there is no drawing involved.

The steel fixers on site require both drawings. The reinforcement drawing tells them where and how to locate the bars, while the bar schedule tells them the shapes of the bars (which is not always evident from the layout).

The steel fixers tie each intersection with soft iron wire. This is one good reason for using the largest bars possible to reduce the number and thus the time and cost of fixing

them, see Fig. 8.8 and Fig. 8.9.

What conventions have been observed in the reinforcement drawing and bar schedule Figs. 8.6 and 8.7?

(a) Reinforcement is shown by single thick, out of scale lines that stand out clearly from all other detail. End views of bars can be made even more out of scale if necessary to show them more clearly.

The Report suggests lines 0.6 mm thick for reinforcement with 0.4 mm for concrete outlines, and 0.2 mm for dimension and centre lines.

(b) Slab dimensions are not shown in views which detail reinforcement.

(c) Location dimensions for the ends of bars are restricted to the section where possible.

(d) Conventional concrete shading is not used on the section. (A light pencil shading on sections, applied to the back of tracings, makes sections stand out on prints).

(e) No attempt is made in the section to separate bars in the same horizontal plane other than by showing hooks. (These are shown as ticks rather than as round hooks).

(f) The first bar in a series is shown whole in plan and a short bit of the last one, but no others. A thin line at right angles to the bars with an arrowhead to the whole bar and the fragment carries a reference, for which the preferred form is: 20R8 - 63 - 150T

which means:

20 bars of round mild steel (R), 8 mm diameter, mark 63 at 150 mm centres in the Top of the slab

The number of bars in each group should be given only once to avoid confusion when totalling on the schedule.

In the section, a bar mark is written parallel to the main reinforcing bars if there is no ambiguity, otherwise an arrow is used to connect the reference uniquely to the bar. The references for longitudinal bars are written normal to the slab, if possible, close to the actual bar or indicated by arrows if necessary.

The bar mark is the unique reference of a bar within the job. On small jobs, the types of bar can be numbered from mark 1 onwards. On larger projects, bars can be numbered sequentially for each structure, e.g., bars for a sluice are marked S1 onwards while bars for a bridge are marked B1 onwards. Bars can also be scheduled for each drawing, so long as each bar in the whole job has a unique reference. This is usually done by including the drawing number in the reference.

The preferred sizes of reinforcement, in millimetres, are 6, 8, 10, 12, 16, 20, 25, 32, and 40.

The types of steel for reinforcement, with their references, are:

R—round, mild steel bars in the metric range of areas.

Y—high yield, high bond strength bars in the metric round range of areas.

X—other than R and Y, for which an explanation is required in the specification and also on the drawings and schedules.

Note the following departure from practice in some offices as stated in the Report:

Although it is common practice in some offices to show top bars by broken lines it is preferable to use full lines for all bars detailed on the drawing, only bars being detailed on another drawing being shown as broken lines.

(c) Standardization

Even this simple example shows what a finicky, time-consuming job the detailing of reinforcement can become. The work is diffi-cult to check moreover and errors in reinforcement cause delays on site.

There are many reasons why reinforced concrete should follow the standardization for rolled steel sections. Here is what the Report says about reinforcement shapes, paragraph one, and standard sections, paragraph two.

BS 1478 defines a standard method of scheduling and provides a set of bar shapes which, in suitable combinations, are suffi-

8.7 *Bar schedules in accordance with BS 1478.*

(a) Graphic representation of bars, which is useful to the steelfixer when looking for the bars.

Member	Bar mark	Type & size	No. of mbrs.	No. in each	Total No.	Length of each bar m	Shape
Wharf Slab	1	R12	1	14	14	199·8	Straight
	2	R12	1	9	9	199·8	Straight
	3	R12	1				(shape: 0·350 / 0·180)
	4	R12	1	4	4	199·8	Straight
	5	R16	1	667	667	4·45	(shape: 4·00)
	6	R16	1	666	666	3·75	(shape: 3·30)
	7	R16	1	667	667	2·95	(shape: 2·50)
	8	R16	1	666	666	1·75	(shape: 1·30)
	9	R16	1	201	201	1·00	(shape: 0·20 / 0·20 / 0·20 / 0·20)
	10	R25	1	401	401	0·30	Straight

BLOGGS & PARTNERS Bar Schedule to B.S. 1478 Drg. Sch. Rev. Ref. Date Site Ref

BLOGGS & PARTNERS Site Ref.		Bar Schedule to B.S. 1478							Drg. Ref. Date	Sch.	Rev.	
Member	Bar mark	Type & size	No. of mbrs	No. in each	Total No.	Length of each bar m	Shape Code	A m	B m	C m	D m	E/R m
Whart Slab	1	R12	1	12	12	199·8	20	199·8				
	2	R12	1	7	7	199·8	20	199·8				
	3	R12	1	1334	1334	1·12	60	0·19	0·36			
	4	R12	1	4	4	199·8	20	199·8				
	5	R16	1	667	667	4·45	33	4·00				
	6	R16	1	666	666	3·75	33	3·30				
	7	R16	1	667	667	2·95	33	2·50				
	8	R16	1	666	666	1·75	33	1·30				
	9	R16	1	201	201	1·00	83	0·20	0·20	0·20	0·20	
	10	R25	1	401	401	0·30	20	0·30				

DOR/G

8.7 (b) Bars defined by the shape code of BS 1478 intended for computerized ordering and costing. The steelfixer will require a reference to know what shape of bar to find and fix.

cient for any problem of reinforcement that can arise. It also provides a form of schedule which is suitable for data processing by the supplier of cut and bent bars.

Offices concerned with repetitive structural work should adopt standard sections for beams, columns, pile caps, etc., and even use pre-printed out-of-scale A4 size drawings of these units requiring only a length and bar size and spacing to be written in by office staff rather than draughtsmen.

(d) Other structural forms

This introduction has been restricted to a simple slab. The Report shows the conventions for beams (where sheer stresses are more critical) and columns and the use of fabric reinforcement (a welded mesh of high tensile steel wire).

Reference.

Brian Broughton: *Reinforced Concrete Detailer's Manual*, Crosby Lockwood, 1969.

8.4 DETAILING STEELWORK

Steelwork is less of a puzzle, in some ways, to the lay observer than reinforced concrete because less of it is hidden. Every bit of a roof truss in a garage or warehouse can be seen and, apart from the deck construction, the details of a bridge can often be taken in as you drive over it.

The famous product of the Industrial Revolution, the iron bridge over the River Severn at Ironbridge, in Coalbrokedale, a few miles from Shrewsbury, is worth a visit. It shows how the craft of the timber joiner was cleverly applied to early construction in iron. Inspection of the roofs of nineteenth century railway stations shows the many different members and sections used by the pioneers of iron and steel.

(a) BS4 and safe load tables

As ideas and design procedures crystallized and steel rolling techniques improved, it became possible, and later economically necessary, to limit and improve the range of standard steel sections. These are scheduled in BS4 from which the examples in Fig. 8.10 are taken, which show the geometric dimensions and properties of the larger universal beams. Note that these are the dimensions and properties in metric units of the standard Imperial sized beams, given in Table 5 of BS4. They are not a range of beams in metric sizes.

BS4 makes no reference to strength. The increase in working strength of structural steel is reflected in the safe load tables published, from time to time, by the British Constructional Steelwork Association. Some manufacturers issue the identical tables between their own covers as promotional literature, and can sometimes be persuaded to give copies to students.

(b) Hollow sections

Several famous railway bridges, erected in the last century, have rivetted hollow sections. The Forth Railway Bridge, for example, has enormous circular hollow sections in compression. Trains crossing the Menai Straits pass through the huge rectangular hollow sections of the Britannia Bridge, while the Saltash Bridge (see page ii) uses large oval sections as tied arches. The solution of the shaping and jointing problems involved in these early uses of hollow sections was expensive. As a result, the hollow section fell from favour until modern gas cutting and automatic welding processes became available.

Small sizes of circular, square, and rectangular hollow sections are mass produced. Tables of properties are available for these, see Fig. 8.11.

Large, hollow elements, frequently trapezoidal in section, can be used to create clear, pleasing structures. The hollow section can be sealed to prevent atmospheric corrosion of the inside while the simple flat outside faces are easy to maintain, see Fig. 8.12.

(c) Composite construction

The high tensile strength of steel is sometimes used to complement the compressive strength of concrete, as shown in Fig. 8.13 and Fig. 8.14. The concrete deck gives some

8.8 *A reinforcing mat of thin bars (about 10 mm in diameter) tied by using a special tool and prepared lengths of soft iron wire. Often plain wire is used and cut to length as needed. The wire is then tied by skilful use of pincers. (Photograph supplied by the Cement and Concrete Association)*

8.9 *The reinforcement for an intricate arrangement of beams with a slab beneath. Each intersection point of the reinforcing bars is tied with soft iron wire.*

B.S. 4 : Part 1 : 1962

B.S. 4 : Part 1 : 1962

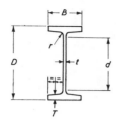

TABLE 5M. UNIVERSAL BEAMS
(Metric equivalents of Table 5)

| DESIGNATION | | Weight per metre | Depth of section D | Width of section B | Thickness | | Root rad. r | Depth between fillets d | Area of section | Moment of inertia | | Radius of gyration | | Elastic modulus | | Plastic modulus | |
Serial size	Weight per foot				Web t	Flange T				About X-X	About Y-Y	About X-X	About Y-Y	About X-X	About Y-Y	About X-X	About Y-Y
in	lb	kg	mm	mm	mm	mm	mm	mm	cm²	cm⁴	cm⁴	cm	cm	cm³	cm³	cm³	cm³
36×16½	260	387	920	420	21·5	36·6	24·1	791	493·9	717 328	42 479	38·1	9·27	15 586	2 021	17 628	3 206
	230	343	911	418	19·4	32·0	24·1	791	436·9	623 868	36 250	37·8	9·11	13 691	1 733	15 445	2 756
36×12	194	289	927	308	19·6	32·0	19·1	819	368·5	503 780	14 793	37·0	6·34	10 874	961	12 556	1 552
	170	253	918	305	17·3	27·9	19·1	819	322·5	435 802	12 512	36·8	6·23	9 490	819	10 930	1 322
	150	223	910	304	15·9	23·9	19·1	819	284·9	375 110	10 424	36·3	6·05	8 241	686	9 505	1 112
33×11½	152	226	851	294	16·1	26·8	17·8	756	288·4	339 130	10 662	34·3	6·08	7 971	726	9 144	1 166
	130	194	841	292	14·7	21·7	17·8	756	246·9	278 833	8 385	33·6	5·83	6 633	574	7 635	929
30×10½	132	196	770	268	15·6	25·4	16·5	681	250·5	239 463	7 701	30·9	5·54	6 223	575	7 156	925
	116	173	762	267	14·3	21·6	16·5	681	220·2	204 747	6 377	30·5	5·38	5 374	478	6 186	773
27×10	114	170	693	256	14·5	23·7	15·2	611	216·3	169 843	6 225	28·0	5·36	4 902	487	5 616	781
	102	152	688	254·5	13·2	21·0	15·2	611	193·6	150 015	5 391	27·8	5·28	4 364	424	4 989	680
	94	140	684	254	12·4	19·0	15·2	611	178·4	135 973	4 789	27·6	5·18	3 979	377	4 552	608
24×12	160	238	633	312	18·6	31·4	16·5	532	303·5	207 252	14 973	26·1	7·02	6 549	961	7 447	1 522
	120	179	617	307	14·1	23·6	16·5	532	227·7	151 313	10 572	25·8	6·81	4 901	689	5 512	1 092
	100	149	610	305	11·9	19·7	16·5	532	189·9	124 342	8 473	25·6	6·68	4 079	556	4 562	884
24×9	94	140	617	230	13·1	22·1	12·7	543	178·2	111 675	4 253	25·0	4·88	3 620	370	4 141	591
	84	125	612	229	11·9	19·6	12·7	543	159·4	98 410	3 676	24·8	4·80	3 217	321	3 672	514
	76	113	607	228	11·2	17·3	12·7	543	144·3	87 262	3 184	24·6	4·70	2 874	279	3 283	449

8.10 Table 5 M of BS 4: 1962.

weather protection to the steel tension element below, while its mass and rigidity both stabilize the structure and transmit lateral loads to the supports.

(d) Fixings

The elements of structional steelwork can be joined, or connected, as the trade says, in four ways. These are, in the historical order of their development, by using bolts, by using rivets, by welding, and by using friction grip bolts.

Bolts. Bolts may be black or high tensile. Black bolts are cheap fixings for temporary attachments (tacking) or lightly loaded parts such as catwalks and ladders. They are black because they are mass produced and remain untreated.

High tensile bolts are made of stronger material with more care taken during manufacture. Neither black bolts nor high tensile ones have any control over the tension or grip they develop on the joint. They must be tight without being overstrained. Their action is almost certain to be in shear (after the completed joint has moved fractionally under load).

Bolts can be turned and fitted (driven often with a large hammer) into accurately reamed holes if movement of a joint must be reduced or almost prevented.

Rivets. Rivets, still used satisfactorily in some shop manufacture, are no longer used on site as the provision of adequate control is difficult and expensive. They are inserted hot into holes, held in alignment by 'tacking' bolts, and are then hammered or squeezed until the plain end is formed into another head. Well fitted rivets almost completely fill their holes and, if made absolutely tight when still very hot, they tighten the joint as they cool and shrink. Their action again is mostly in shear with perhaps less movement of the joint than with bolts because the rivet should fill the clearance in its hole.

CIRCULAR HOLLOW SECTIONS
HOT FINISHED SEAMLESS
High Yield Stress Steel to B.S.968 : 1962

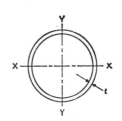

Outside diameter in.		$6\frac{5}{8}$						
Thickness S.W.G. or in.		7	6	5	$\frac{1}{4}$	$\frac{9}{32}$	$\frac{5}{16}$	$\frac{3}{8}$
W	lb./ft.	12·2	13·2	14·6	17·0	19·0	21·2	25·0
Area	sq. in.	3·57	3·88	4·27	5·01	5·60	6·20	7·36
I	in.⁴	18·6	20·1	22·0	25·5	28·3	30·9	36·1
Z	in.³	5·60	6·06	6·64	7·69	8·53	9·34	10·9
S	in.³	7·32	7·95	8·72	10·2	11·3	12·5	14·7
r	in.	2·28	2·28	2·27	2·26	2·24	2·24	2·21
Axial tension-tons		48·20	52·38	57·64	67·64	75·60	83·70	99·36
	1	48·20	52·38	57·64	67·64	75·60	83·70	99·36
	2	48·14	52·32	57·57	67·55	75·50	83·59	99·22
	3	47·93	52·09	57·32	67·25	75·16	83·21	98·75
	4	47·61	51·75	56·94	66·80	74·65	82·64	98·06
	5	47·18	51·28	56·42	66·19	73·94	81·87	97·11
	6	46·62	50·66	55·73	65·37	73·01	80·83	95·84
	7	45·87	49·85	54·83	64·30	71·79	79·48	94·17
	8	44·91	48·81	53·67	62·92	70·20	77·72	92·00
	9	43·68	47·47	52·18	61·14	68·16	75·46	89·21
	10	42·13	45·78	50·29	58·90	65·59	72·61	85·68
	11	40·21	43·70	47·97	56·14	62·42	69·11	81·35
	12	37·94	41·24	45·23	52·88	58·69	64·98	76·28
	13	35·39	38·47	42·15	49·24	54·55	60·39	70·69
	14	32·69	35·53	38·90	45·41	50·22	55·60	64·92
	15	29·99	32·59	35·66	41·59	45·93	50·85	59·25
	16	27·40	29·78	32·56	37·96	41·87	46·35	53·92
	17	24·99	27·16	29·69	34·59	38·12	42·20	49·03
	18	22·80	24·78	27·07	31·53	34·73	38·45	44·62
	19	20·82	22·63	24·72	28·78	31·68	35·08	40·68
	20	19·05	20·71	22·61	26·33	28·97	32·07	37·17
	21	17·47	18·99	20·74	24·14	26·55	29·40	34·06
	22	16·07	17·46	19·06	22·19	24·40	27·02	31·29
	23	14·81	16·10	17·57	20·45	22·49	24·90	28·83
	24	13·69	14·88	16·24	18·90	20·78	23·00	26·63
	25	12·68	13·79	15·05	17·51	19·25	21·31	24·66
	26	11·78	12·80	13·97	16·26	17·87	19·79	22·90
	27	10·97	11·92	13·01	15·14	16·63	18·42	21·31
	28	10·23	11·12	12·14	14·12	15·52	17·18	19·88
	29	9·57	10·40	11·35	13·20	14·51	16·06	18·58
	30	8·97	9·74	10·63	12·37	13·59	15·05	17·40

(First left column, rotated: Axial compression in tons for effective lengths in feet.)

8.11 *Representative tables of properties of hollow steel sections. (From* Structural Hollow Sections, *Stewarts and Lloyds Ltd.)*

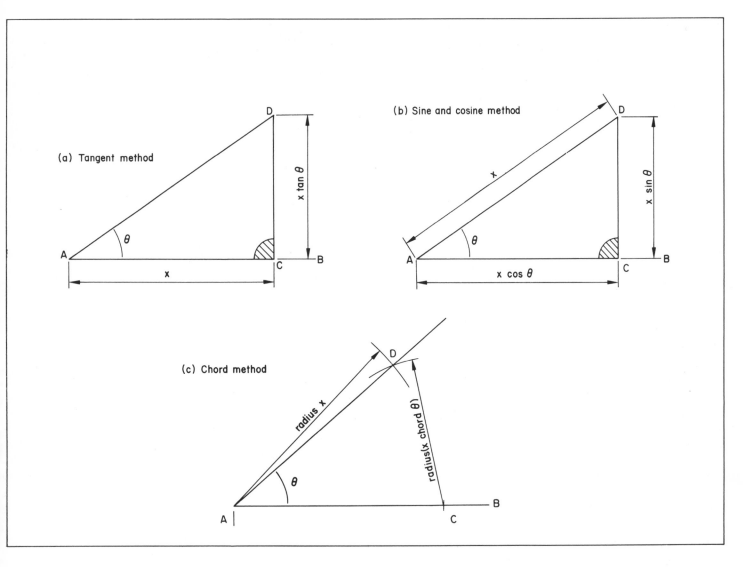

(a) Tangent method

(b) Sine and cosine method

(c) Chord method

Swing arc centre A radius x units.

Centre C, swing an arc radius x times the chord of angle θ.

Join AD.

Method (c) is the quickest and most accurate method if a table of chords is available.

Of course the acute angles between the traverse lines at the survey station should be used for greater accuracy and convenience.

Co-ordinates. Because of the likelihood of errors from plotting traverse lines with a protractor or even the more accurate methods of setting out angles, as shown in Fig. 8.19, survey stations, which have been located very accurately by triangulation with a theodolite, are located by co-ordinates called latitudes and departures (or eastings and northings). The co-ordinates of each station, relative to a pair of arbitrarily chosen axes at right angles, are calculated by means of five or seven figure logarithms, calculating machine, or computer.

Each survey station is plotted with reference to a grid which is set out on the plotting sheet with meticulous care and accuracy, see Fig. 8.20. Here are the instructions for plotting a grid:

(a) Draw any convenient line AB with a straightedge.

(b) Choose point C and erect a perpendicular, using the method described below.

(c) Set off CD a convenient multiple of grid units.

(d) Set off CF a convenient multiple of grid units.

(e) Centre D, radius CF, swing an arc DE.

(f) Centre F, radius CD, swing an arc FE to cut the first arc at E.

(g) Divide CD and FE from C and F into grid units 1, 2, 3, etc., and join with the straightedge.

8.19 Setting out angles.

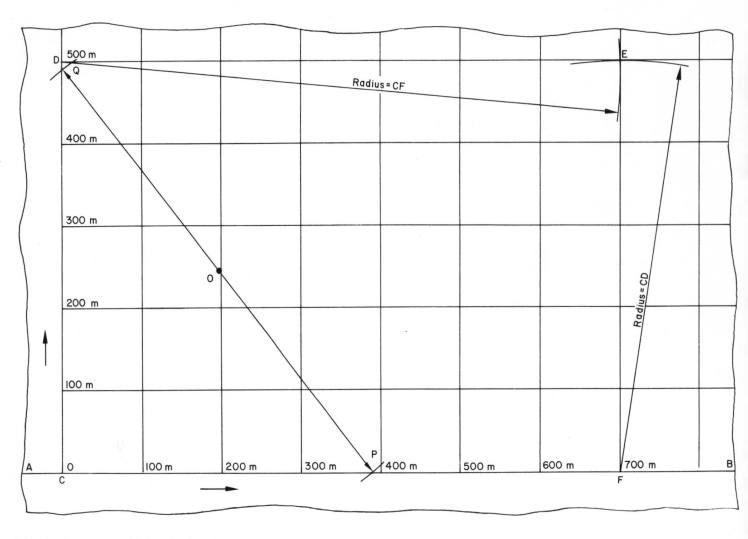

8.20 *Setting out a grid for plotting the co-*
      ordinates of survey stations, and how to
      set out a right angle.

(h) Similarly divide CF and DE and com-
     plete the grid.

The gridding operation should be completed
as soon as possible to prevent errors due to
paper instability.

Survey stations are located from the nearest
grid lines, not from the axes.

The temptation to save time and plot co-
ordinates without gridding but by measurements from the main axes should be resisted.

Here is the method of erecting a perpendicular through a chosen point C on a line AB,
as shown in Fig. 8.20.

(a) Choose any point O.
(b) Centre O, radius OC, swing an arc to
     cut the given line at P.
(c) Produce the line PO to meet the other
     side of the arc at Q thereby drawing
     the diameter POQ.

The angle PCQ is the angle in a semicircle
and is therefore a right angle, hence CQ is
perpendicular to the given line.

The arc QCP should be as large as possible

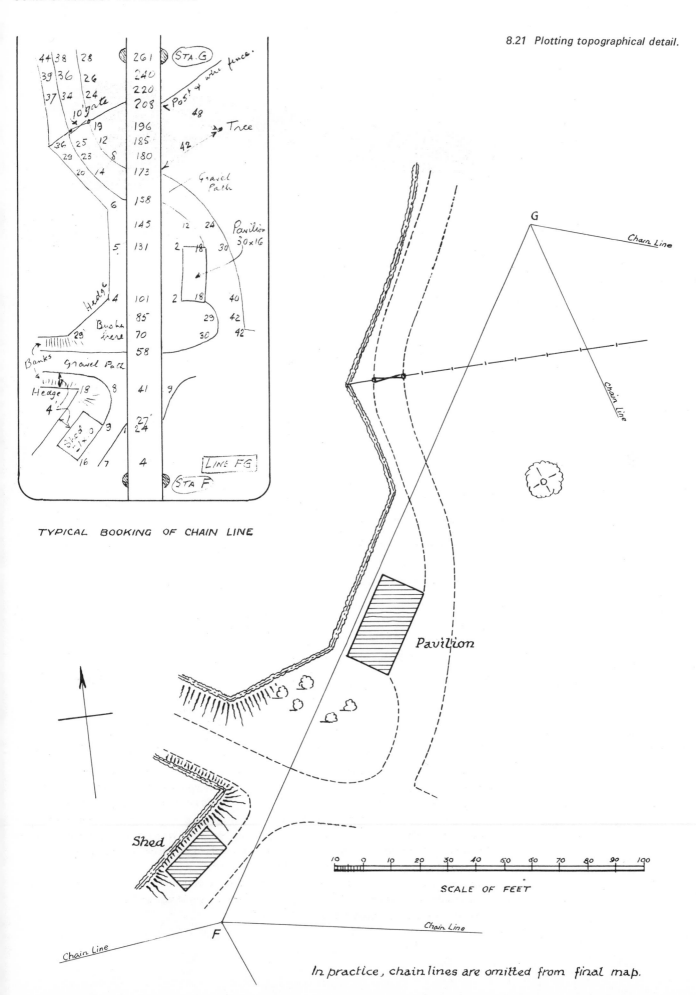

8.21 *Plotting topographical detail.*

TYPICAL BOOKING OF CHAIN LINE

SCALE OF FEET

In practice, chain lines are omitted from final map.

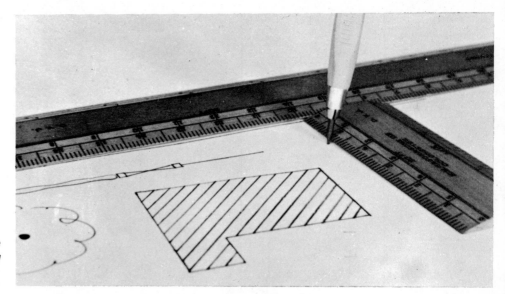

8.22 *Using an offset scale for quick, accurate plotting of offset measurements normal to a chain line.*

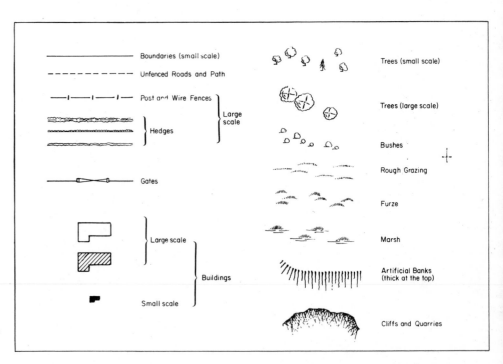

8.23 *Some common conventions for use when plotting surveys.*

and must therefore be swung with beam compasses.

(c) Filling in topographical detail

By whatever method the traverse has been surveyed and plotted, most of the topographical detail will be located by simple, offset measurements from the traverse lines. Most offset measurements will be made at right angles to the traverse line.

Figure 8.21 shows a page from a field book

on which offsets and chainage are recorded together with a plot of the same detail.

An offset scale speeds plotting, see Fig. 8.22. It is a short scale, fully divided to its square ends, which can slide along a full size scale held in place on the traverse line by weights.

Some common conventions used in plotting surveys are shown in Fig. 8.23.

Chapter 9

The drawing office and the reproduction of drawings

Engineering on a small scale: Kingsgate footbridge at Durham. To save erecting scaffolding in the river, the symmetrical halves of the bridge were built parallel to the banks and, on completion, were swung out to meet each other. The engineer was given a very free brief for this bridge and devised this novel, but very elegant and economical, solution. (Photograph supplied by the engineer, Mr Ove Arup of Ove Arup & Partners)

9.1 INTRODUCTION

In many establishments, the distinction between the design office and the drawing office (DO) is becoming less distinct. Many finished drawings are now prepared by graduate engineers working in the design office because the number of skilled draughtsmen and tracers is steadily dwindling and their services are almost prohibitively expensive.

Many new engineers start their professional lives as part of a design team, spending much of their time at a drawing board, if not actually in a drawing office.

This chapter is concerned with the routine side of the DO, and is an introduction to referencing, filing, and storing drawings, and also to methods of reproducing them.

Traditionally, engineering drawings have been made on cartridge paper and reproduced full size from an ink tracing by a semipermanent photographic process known as dyeline or diazo.

The sheer bulk of full size drawings, tracings, and prints creates storage problems. The solution to these problems is microfilming by which a transparent photographic copy, small enough to be mounted in a hole in a standard computer card, is made. Enlarged prints can then be produced by projection of the transparency.

9.2 PLAN REFERENCES

Each drawing must have its own reference number.

Smaller firms may simply number each drawing as it is produced, irrespective of the scheme to which it belongs. On the other hand, they may follow the practice of larger organizations, by which each scheme has a reference and all drawings for that scheme are consecutively numbered under that reference.

A simple system must be devised that will prevent more than one drawing being issued with the same number. A single register, with the drawing numbers already written in, could be used. It is surprising how many errors can occur if each person who registers a drawing has to write down a number following on consecutively from the one above. A minimum of information, the title of the drawing, its size, and the date of registration, is entered against each number.

If more information is required, card index systems can be used to contain some or all

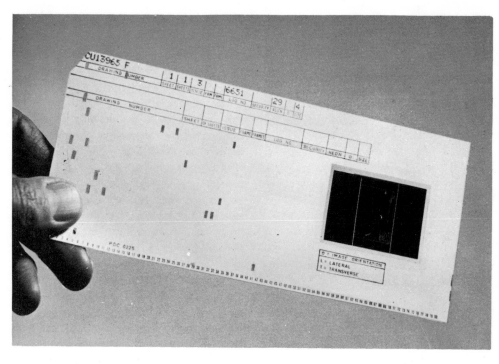

9.1 *An aperture card.*

of the following items, or even more:

(a) drawing number
(b) title
(c) date registered
(d) size
(e) details and dates of amendments
(f) whether drawn on paper or film
(g) whether traced
(h) whether microfilmed
(j) to whom issued and how, i.e., full size print, reduced print, microfilm copy, or intermediate print
(k) whether original kept or destroyed
(l) whether in current use or in archives
(m) whether completion or 'as-built' copy prepared.

An organization must decide how much information it wants to record, and how it is going to keep the records up to date. A registry clerk will keep records better than an engineer and will ensure that engineers provide the necessary data. A simple system, properly kept, is better than an over ambitious one that is incomplete.

9.3 STORAGE

Microfilms, see section 9.5, mounted in standard computer cards, aperture cards, as shown in Fig. 9.1, are easily stored in small drawers or boxes, depending on the number involved. Some large concerns sort and retrieve these cards by using mechanical systems.

Even if an office records its drawings on

9.2   *Horizontal storage in drawers. (Admel International Ltd.)*

9.3   *Vertical storage, a suspension system. (Admel International Ltd.)*

microfilms, some storage for full size drawings will be needed; for the working prints taken from the microfilms (assuming they are returnable) and, certainly, for the original in case of amendments.

Any system of storage must give quick and easy access to any drawing it contains. Drawings may be stored horizontally in drawers, see Fig. 9.2, or in one of the vertical filing systems, see Fig. 9.3.

Drawers more than 50 mm deep become very heavy when full, while replacing drawings in a full drawer or filing a flimsy tracing is almost impossible, with or without help. Horizontal storage chests require at least twice their own floor area for convenient operation. The smaller units of storage provided by each drawer, however, make these storage chests popular.

Vertical storage systems occupy less floor space than horizontal chests for similar storage and need considerably less space in operation. Each system of vertical storage boasts that any drawing can be located quickly and be taken out and put back easily.

The differences in the systems result from the ingenious methods devised to prevent sheaves of slippery drawings and flimsy tracings from slipping to the bottom of the cabinet, while at the same time, maintaining easy access to all drawings. Some systems suspend the drawings, some grip them very firmly, and one corrugates them to stiffen them so that they do not collapse to the bottom of the cabinet, see Fig. 9.4.

A strong perforated strip must be attached to one edge of each drawing in the suspension systems. These strips are not cheap and take time to fix. Such systems, however, can cope with mixed sizes of drawings in one cabinet.

Small or odd size drawings upset most systems; they slide to the back of drawers and need special provision in the vertical systems unless they are suspended.

9.4 REPRODUCING DRAWINGS

There are two main methods in use for reproducing engineering drawings, the dyeline process and the electrophotographic process. The dyeline process produces a black line from a black line in the original, unlike the normal photographic process where an intermediate, negative, stage is introduced. The electrophotographic or electrostatic process

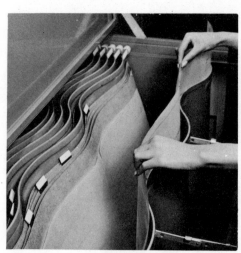

9.4   *Vertical storage in which drawings are corrugated. (Roneo-Vickers Planfile)*

can be set up to make black line prints from positive or negative originals.

In the dyeline process, a sensitized coating on the paper is developed, after exposure to light, to produce a dark line of dye where there was a line on the original. Development is either dry by ammonia gas or semidry when the coated surface is dampened with a minimum of developing solution that soon soaks into the paper or evaporates. The image is not fixed in the photographic sense and deteriorates in a few weeks if continuously exposed to daylight. It is almost permanent if prints are kept in the dark when not in use. Exposure and developing can be done in the same machine, see Fig. 9.5.

Originally dyeline prints were always made the same size as the original tracing by exposing the paper to light through the tracing which was in contact with it. Prints can now be made by projection from microfilm positives (see below) on to dyeline paper thus making possible almost any degree of enlargement or reduction from the original size.

Reductions can be made without microfilming by photographing the original to the size required with very large cameras and making a positive transparency for dyeline printing from the negative. A positive is necessary for each degree of reduction.

In the electrophotographic process, an image of the original is projected on to an electrostatically charged photoconductive material. The action of light causes the electrostatic charge to leak away leaving lines of charge in place of black lines on the original. A reversal of the charge makes it possible to reverse the image and print directly from the negative image obtained in the microfilm camera.

The photoconductive material can be part of the machine, often a selenium coated drum, or a coated paper with a conductive base.

Fine carbon particles, attracted by the electrostatic charge, are transferred from the drum to the printing paper and are fixed by heat fusion, or they are attracted directly to the conductive based paper and then fused. If the fusion is complete, the image is virtually permanent, otherwise the carbon dust rubs off almost at once or in the post, which can lead to confusion.

Before the advent of polyester film, all tracings were made on specially treated fine linen. The linen is either dyed blue or left white, neither leaves any background on the print as dyeline paper does not react to blue light. For this reason, blue pencil lines, used to lay out a drawing, need not be erased.

Polyester film is very stable, does not react to damp as linen does, and is more permanent. It even resists the attacks of tropical ants.

The traditional method of reproducing drawings, full size on dyeline paper through a tracing from an original made on cartridge paper, is still used and produces superb work. The tracing process is slow and expensive but it is also flexible, e.g., views can be rearranged, etc. Dyeline prints, however, are cheaper than any other form of reproduction.

Today, most drawings are made directly on to polyester film or tracing paper from which good prints can be made of pencil work. Some offices ink in dimensions or background material for greater clarity.

Economic pressures and the demand for greater output have forced engineers to accept the inevitably lower standard of work brought about by omitting the tracing stage. Tracers transform sketchy and rough drawings, they tidy up and space out confused views, and their printing is immaculate. The engineer himself must now locate the views correctly and print legibly.

Dyeline prints can be made on a transparent base. These are called intermediates and print as well as the original. If very many prints are required, it is sensible to produce intermediates for use in several offices. The intermediate process is also used when alternative arrangements of schemes are to be fitted to the same background. Several intermediates are made of the background and are then finished separately in ink or pencil.

The polyester film is glossy on both sides when it is made. The rough surface necessary for pencil or ink drawing is usually achieved by coating the glossy surface, but some firms roughen the surface mechanically. Both types of surface break down with persistent rubbing out and leave a shiny surface which will take neither ink nor pencil.

9.5 MICROFILMING

A microfilm is a very small photograph, on transparent material, of an original drawing. The photograph must be enlarged for use.

Microfilming requires a substantial outlay on equipment, and is undertaken only under pressure by most organizations. The two main pressures are:

(a) Shortage of storage space.

(b) The need for quick access to up to date drawings.

Early advocates and developers of the process were hopelessly overoptimistic, or ahead of photographic emulsions and lens design, and started working on 16 mm film which is far too small for drawings although suitable for some document work. After some work with larger sizes (70 and 105 mm film), routine microfilming now uses unsprocketed 35 mm film which gives a usable area of about 30 x 45 mm.

A wide variety of systems is available but, in general, drawings on cartridge paper or a translucent base are photographed on to a

9.5 Medium capacity, combined printer-developer. It is used for reproducing drawings by the dyeline process from originals on translucent material. The original and paper are fed in together, the paper reappears after exposure and is fed straight back for semi-dry development. (Monex 400, Admel International Ltd.)

9.6   *A processor camera that makes a 35 mm microfilm ready mounted in an aperture card from a 1000 mm x 760 mm original in 40 seconds. (2000 Processor Camera by 3 M)*

normal silver halide film to give standard photographic negatives, see Fig. 9.6.

If prints are to be made on dyeline paper, a positive transparency must be made, otherwise negative prints will result which are not very convenient to use. You cannot sketch or write notes on black paper; and engineers always sketch or write notes on drawings.

As the electrostatic process can reverse the image, regeneration can take place directly from the original negative.

Negative or positive transparencies are mounted in aperture cards. Most offices are used to handling such cards and have storage facilities or even mechanical sorting and retrieval systems.

Details of the drawing can be typed and punched on to the card.

Prints up to 1000 x 750 mm can be made, but 600 x 450 mm is more common. The latter amounts to about 2:1 linear reduction from an A1 drawing (a quarter of the area), see Fig. 9.7.

Original drawings must be clear without too much fine detail or small print. Letters less than about 4 mm high do not reproduce well. Equally important is the thickness of

the lines. A well proportioned small symbol or letter will reproduce better than a large but spidery one.

The archival advantage of microfilming is obvious, the vast bulk of original drawings, tracings, etc., can at least be halved by making the tracings obsolete. Original drawings must be kept for amendments until the job is completed.

Microfilm is used in viewers or enlarged prints are produced.

An engineer, with a viewer at his desk and a compact store of aperture cards, can refer to any drawing at once without having to search through vast quantities of paper.

Much time can be spent in collecting drawings from a central filing system, chasing up the ones already out, and refiling them after use. Some offices have overcome this problem by issuing non-returnable microfilm prints. The cost of the system is covered by the saving in time. The central store of microfilm negatives is easily kept up to date, while the traffic in drawings is smooth and in one direction, outwards. One installation can have a drawing in service in less than five minutes after receiving an order.

9.7   *A reader-printer that makes electrostatic prints 457 mm x 609 mm in 25 seconds from 35 mm microfilm mounted in aperture cards or throws a viewing image on a screen 465 mm x 650 mm. (Ricoh International, Tokyo)*